Oxford Physics Series

General Editors

E. J. BURGE   D. J. E. INGRAM   J. A. D. MATTHEW

Oxford Physics Series

B. R. JENNINGS and **V. J. MORRIS**

READER IN PHYSICS
BRUNEL UNIVERSITY
UXBRIDGE

RESEARCH FELLOW IN PHYSICS
BRUNEL UNIVERSITY
UXBRIDGE

Atoms in contact

Clarendon Press · Oxford · 1974

Oxford University Press, Ely House, London W.1

GLASGOW   NEW YORK   TORONTO   MELBOURNE   WELLINGTON
CAPE TOWN   IBADAN   NAIROBI   DAR ES SALAAM   LUSAKA   ADDIS ABABA
DELHI   BOMBAY   CALCUTTA   MADRAS   KARACHI   LAHORE   DACCA
KUALA LUMPUR   SINGAPORE   HONG KONG   TOKYO

PAPERBACK ISBN 0 19 8518048

CASEBOUND ISBN 0 19 8518099

© OXFORD UNIVERSITY PRESS 1974

PRINTED IN GREAT BRITAIN BY
J. W. ARROWSMITH LTD., BRISTOL, ENGLAND

Editor's Foreword

TRADITIONALLY students of physics begin with a macroscopic view of the properties of matter, and, after serving some apprenticeship in the intricacies of atomic and quantum physics, they graduate to a study of the microscopic picture of the liquid and solid state. *Atoms in contact* presents the atomistic view, but asks no more than basic school physics as a background. The result is a sustained essay which conveys a vivid impression of the diversity and subtlety of the behaviour of atoms in close interaction. The book can form the basis of an elementary course for physics students, but may also serve as an introduction to solids and liquids for electrical engineers, chemists, and materials scientists.

The treatment in this text fits carefully into the integrated structure of the Oxford Physics Series. Complementary core texts cover *Radiation and quantum physics*, *Atoms and their structure*, *Mechanics and motion*, *Interactions of particles*, *d.c. and a.c. circuits*, and *Electromagnetism*, and pave the way for more advanced books. *Atoms in contact* in conjunction with the background material in the other core books leads naturally to a more detailed and more advanced view of condensed matter in *The solid state* by H. M. Rosenberg. The other books in the series can similarly be used in a flexible way to fit diverse styles of physics courses, and yet give an integrated and coherent picture of physics.

J.A.D.M.

Preface

So often in the past, books have been written on the properties of matter in which the subject has been presented predominantly from a macroscopic viewpoint and which have dealt mostly with the mechanical properties of materials. The aim of the present text is to give a cursory view of some of the important characteristics of solids and liquids through a wider realm of physical properties.

Our approach is from the atomic standpoint. By considering the basic properties of atoms themselves and of their interatomic forces when they are brought into close proximity, we try to show how the materials so formed, be they orderly crystals or the somewhat more random glasses or liquids, depend both upon the basic properties of the constituent atoms and on the conditions prevailing when they were brought together. It is not surprising that the atoms indelibly stamp their nature on the solids or liquids they form. We are familiar with the way that a random, roaring crowd at a football match has a character of its own which is very different from that formed by the same folk when they are in their orderly array in church on Sunday; yet, just as the people are the same in both aspects of this analogy, so are the atoms in the various states of condensed matter.

Much of our current, conventional teaching of physics compartmentalizes concepts and ideas. Various models are used to explain various behaviours. Examples which touch upon the subject matter of this book are the use of *chemical bonding theory* to explain how pairs of atoms are held together. Yet when dealing with solid-state theory and the electrical properties of solids, the student is presented with a completely different approach in terms of the *band theory*. Other effects are explained in terms of *quantum theory*. The student is expected to have the mind of a grasshopper. Yet the same atoms and materials are the basis of all the properties, and no single approach has a monopoly of the 'truth'. Rather, each emphasizes one factor which makes it more appropriate to specific physical effects. In this text, therefore, we have tried to show how the various models are interrelated, and applied to account for the observable properties of materials.

The reader envisaged is the student in transition from A-level to a first year undergraduate course: the text could form part of a foundation course in physics. We hope it will prove more than useful to anyone else who wishes to gain a broad sweep of the properties of materials. To this end, above all else, the authors have tried to make the text *readable*. It is our hope that a conscientious, intelligent student can read the book in a few evenings and gain a conceptual view of the properties of solids and liquids. We have deliberately tried to avoid the use of jargon, mathematics, and problems, in the hope that the reader may gain a feel of the subject without getting lost in the details. This is not to say that we believe that a competent student of physics need not appreciate the theoretical treatments of his subject; he must, and mathematics should be one of the major weapons in his armoury. Nevertheless, we feel that there is a need for a text which explains the phenomena simply.

Such a work cannot be complete. The restricted length of the volume has caused us to reject certain subjects which we would have liked to have included. Also, it would be fair to say that, of the twenty-two subheadings used in this book, each in its own right has been the subject of books of equal or greater length than this volume. For this reason we cannot be rigorous; but we have tried to provide a large and helpful bibliography where each topic is more justly treated.

Finally, it is a pleasure to acknowledge the interest and encouragement of Dr. J. A. D. Matthew of York University. We are especially indebted to those who have read the text in manuscript form and given us advice. We count Dr. J. A. D. Matthew, Dr. J. Warren, Mr. R. Hall and Dr. S. Konidaris (the last three mentioned of this University) as worthy of special thanks. Our gratitude is also expressed to Mrs. J. Coles for typing the manuscript, to Dr. C. Elam and the O.U.P. for permission to reproduce Fig. 3.5, to Professor A. Forty and Dr. H. F. Kay of Warwick and Bristol Universities respectively for providing the photograph of Fig. 2.5, and to Professor P. Feltham of Brunel University for supplying Fig. 2.6.

Brunel University   B. R. JENNINGS
                                                               V. J. MORRIS

Contents

1. What is condensed matter?

'He had a way of meeting a simple question with a compound answer—you could take the part you wanted, and leave the rest.'

RING LARDNER *Dinner bridge*

1.1. Atomic structure of matter

THE building blocks of matter are atoms. These are discrete particles which are associated together in various ways and to varying degrees to constitute a given material. Although the particulate nature of matter was anticipated by the Ancient Greeks, atomic theory became popular in the seventeenth century in explaining the properties of solids. Today, we know that atoms are approximately spherical with diameters of the order of 10^{-10} m. They consist of a minute, extremely dense, and positively charged nucleus, which is itself composed of heavy, uncharged neutrons and positively charged protons of nearly equal mass (of the order of 1.6×10^{-24} g). The nucleus is surrounded by a cloud of relatively light (approximately 10^{-27} g), negatively charged electrons. If viewed as particles, the electrons are most likely to be found in certain volumes of space, governed by their so called *atomic orbitals*. The occupancy of the orbitals obeys strict exclusion rules (see Chapter 6). As isolated atoms are electrically neutral, the number of electrons is equal to the number of protons in the nucleus. The chemical properties of atoms, including their combination to form molecules, are governed predominantly by the outermost electrons which are least strongly bound to the nucleus and hence externally accessible.

Now atoms may associate and form a specific material; but even this material can assume a variety of forms. For example, if an ice cube is held in the hand, it absorbs energy in the form of heat, and melts to a puddle of water. Holding one's hand in front of a fire provides further energy and the water may evaporate as vapour. Ice (a solid), water (a liquid), and water vapour (a gas)† are examples of three common states of matter. During a change of state the chemical composition of the material is unaltered. Ice, water, and steam are all composed of H_2O molecules (a stable arrangement of hydrogen and oxygen atoms). Why then are the physical properties of the various states so different? To answer the question we must consider the forces between atoms. These are of two types: namely attractive and repulsive. Both types of force are short range but, as the interatomic distance is increased, the magnitude of the repulsive force drops more rapidly than that of the attractive force (Fig. 1.1).

† The term *vapour* is used for a gas when below its so called *critical temperature*.

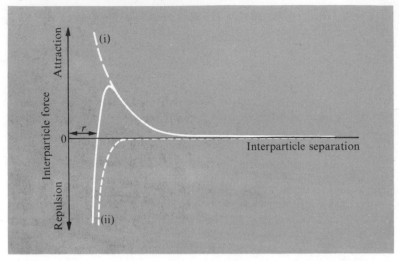

FIG. 1.1. Interparticle-force curve for two idealized atoms. The attractive force (i) is relatively long range; the repulsive force (ii) is not. The full curve represents their resultant. Zero force is experienced when the particles are separated by distance r or infinity.

1.2. States of matter

What happens when the atoms of a substance are distributed within a volume very large in comparison to their own size? Do they congregate at one point or spread evenly throughout the volume? The key lies in the net internal energy possessed by the material. When the internal energy is large, the atoms are not stationary but move in straight lines at random velocities between chance collisions with the atoms of the container or those of the material. Under such conditions the motion of the atoms, and hence the shape of the material, is restricted by the shape and size of the container. Such a medium is said to be in a *gaseous state*, and the motion of the atoms, and hence the overall properties, are essentially independent of the constituent atoms or molecules. In such a state the mean interatomic separation is dictated by the container volume and may be many times the atomic diameter.

Suppose we reduce the internal energy (i.e. lower the temperature). We may do this by allowing atoms which collide with those of the container to lose energy. We visualize an inelastic collision in which the atoms rebound with velocities of lower magnitude than those with which they approached the wall. Collisions between atoms of the material are still on average elastic. Some of the atoms may adopt a more compact and more ordered arrangement in which the interatomic separation, and hence their overall volume, is reduced. They may still move about, but they are now so restricted that their motion is no longer independent. The properties of the material now depend increasingly more on the nature of its constituents and their characteristic interatomic forces than on the size of the container. The material has condensed from the gaseous into the *liquid state*. Although the atoms are in close proximity to each other, their freedom of motion still permits the material to flow. Like a gas, the liquid is thus fluid. This term simply means 'ability to flow' and should not be confused with 'liquid'.

Should the liquid be further cooled, the energy of the atoms continues to fall. Ultimately, the material loses its fluidity and the interparticle forces become all important. The atoms are drawn into somewhat definite closely packed positions. Internal energy simply causes the atoms to oscillate about these positions. The material becomes hard, and maintains a definite shape. The liquid has frozen into a *solid state*. Under certain conditions, the atoms may adopt very specific relative positions and collectively form a stable, three dimensional array. This most ordered array is a *crystal* and the binding forces are termed *bonds*. The crystal structure is thus a manifestation of the cohesive forces which in turn are characteristic of the atomic structure. Unlike the atomic positioning in gases, the atomic array in solids is very specifically related to the structure of the constituent atoms.

1.3. Condensed matter

The transition from gas to solid may thus be considered as a direct consequence of the net internal energy of the atoms competing against the ordering effect of the interatomic forces. At the gaseous extreme, disorder predominates and the atomic structure is relatively unimportant for the physical behaviour. With solids, the atomic structure and bonding are all important. It is this ability of the atomic nature to assume importance that gives rise to the wide range of properties of different solids. Hence, it is convenient to use the degree of disorder as a measure of the division of materials between gases and the various condensed states of matter, i.e. liquids and solids. In the latter class, the atoms are in close enough proximity to be considered as 'in contact' and it is with this class that this book is concerned.

The high density of many metals plus their discrete and sharp edges, suggests that the interparticle forces are strong in this class of material and that the atoms are in a regular, close-packed lattice. In contrast, the more open atomic structure in liquids allows these substances to be mobile and to adopt the shape of a container. Such straightforward properties are easily envisaged in terms of the atomic order in condensed matter as presented above. Yet is this interatomic picture able to account for other more complicated properties of solids and liquids? Can it explain why certain solids are strong, whilst others are brittle, or why some are rough, smooth or greasy to the touch? Does it give us the key as to why graphite is a good electrical conductor, whilst diamond is a poor one; even though they are both forms of carbon? Does it indicate why certain materials can be made to emit light without light being shone onto them, or why the electrical properties of silicon and germanium are extremely sensitive to parts per million of certain impurities whilst other materials can tolerate up to ten per cent of impurity with little change in their properties? Alternatively, we might consider changing or modifying the atomic binding energies in order to obtain different materials. Is it possible to make a solid which does not have the fully ordered crystal lattice, but has the less ordered atomic array typical of a liquid? If so, how will the properties of this material differ from those of the crystalline solid?

These are typical of the fascinating properties that are controlled directly or indirectly by the interatomic bonding which is brought into play when atoms are forced into close proximity as in the condensed states of matter. In the following chapters an attempt is made to show how the individual nature of atoms and the effect of the *entente cordiale* (or otherwise—*vive la différence*) between them accounts for many of the physical properties of solids and liquids.

2. Condensed matter described

'They (atoms) move in the void and catching each other up jostle together, and some recoil in any direction that may chance, and others become entangled with one another in various degrees according to the symmetry of their shapes and sizes and positions and order, and they remain together and thus the coming into being of composite things is effected.'

SIMPLICIUS (530 B.C.) *De caelo*

2.1. Interatomic forces

IN its various forms, condensed matter is very common in nature. It can neither be infinitely expanded (without becoming gaseous) nor indefinitely compressed. These elementary properties suggest an interplay of attractive and repulsive forces between the atomic constituents. The physical and chemical properties of the material depend on the constituent elements. We must ask ourselves how the interatomic forces arise and why they depend on the initial atomic structure.

The key to an understanding clearly lies in our picture of the structure of an isolated atom. We imagine a positive nucleus surrounded by electrons. These electrons are constrained to certain volumes in space, governed by their *atomic orbitals*. With condensed matter, the constituent atoms are only a few angstroms apart (1 Å $\equiv 10^{-10}$ metres), so that we are dealing with atomic densities of the order of 10^{28} to 10^{29} atoms/m³. Under such conditions, the orbitals overlap and the outer electrons are no longer exclusively owned by individual atoms but are cooperatively shared by the whole material. Thus, in a crystal, for example, one must then think of the electrons as confined to *crystal orbitals* rather than their original atomic orbitals. However an exact description of these orbitals is not possible at the present time owing to the large number of interacting particles (electrons and nuclei) to be considered. Because of this, approximate models must be used to discuss the electronic structure of condensed media.

In one approximation, the nuclei of the system are considered in their equilibrium positions. The bonding between the atoms is then considered as arising from the forces which accompany the distortion of the atomic states as the atomic orbitals overlap. This approach is known as the *Linear Combination of Atomic Orbitals* and is an extension of *Molecular Orbital* theory. It is attractive in that it accounts for the effect known as *hybridization*, where more than one type of atomic orbital per atom is involved in the interaction. This complexity usually dictates the spatial orientations of the atoms in the medium and generally results in very strong bonding. The alternative approach to crystal orbitals is to consider those atomic orbitals of the equivalent, isolated atoms which associate to form the bonds experienced in the condensed state. That is, we choose arrangements of electrons and nuclei which seem most likely to occur and omit arrangements which are physically unlikely.

The method is often called the *Valence Bond* approach. In practice, both models need to be adjusted to produce realistic representations. However, to a first approximation each suggests that the crystal orbitals can be pictured as arising from bonds between pairs or groups of atoms.

So far we have said that as atoms come close together, the electronic motion is perturbed and bonds are formed. We have said little about how great this perturbation is, or why it occurs. Condensed matter is formed when elements react together, liberating energy and forming a state of matter which is energetically more favourable than that of the isolated atoms. Chemists noticed that the group O elements,† the inert gases, showed a great reluctance to react with themselves, or, indeed, with any other element. They attributed this stability to the unique structure of the outer electronic shell (called the *valence shell*), whose occupied states were filled to capacity. It was proposed that an atom could be treated as a positive unit composed of the nucleus, surrounded by a core of inner electrons, which in turn is encompassed by the negatively charged valence shell. It was then suggested that chemical reactions basically involved changes in the distribution of valence electrons, with each reactant endeavouring to obtain a stable inert-gas-like electronic structure. The types of electronic transitions which characterized the early ideas of chemical bonds are outlined below.

2.1.1. *Ionic bonding*

Ionic bonding involves the exchange of electrons by the participant atoms. For example, the exchange of one electron from a group I element to a group VIIB element results in a positively charged *cation* and a negatively charged *anion* (Fig. 2.1(a)). Each of these then possesses an inert-gas-like electronic structure. In a resulting condensed medium the attractive forces are due to electrostatic interactions between these cations and anions. There are two sources of repulsive force. Firstly, electrostatic repulsion exists between like ions. Secondly, if the ions are forced too closely together, the inner-core electronic orbitals interact. These orbitals are already filled with electrons so that the interaction is equivalent to an attempt to insert extra electrons into them. The result is a strong repulsive force termed *inner-core repulsion*.‡

2.1.2. *Covalent bonding*

Many materials are non-ionic. To account for bonding in such materials the idea of sharing electrons was introduced. Consider a group IVB element. In order to obtain an inert structure the atom must gain four electrons. However, by gaining a single electron the atom becomes negatively charged, and the resultant anion strongly resists the addition of further electrons. To add four electrons would be a formidable task indeed. However, if each atom

† The reader is referred to the periodic table at the end of the book.

‡ Repulsion arises because the occupancy of energy levels is limited. No more than the allowed number of electrons may be forced into a given level. See chapter 6.

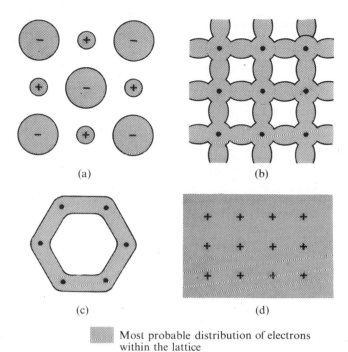

F IG. 2.1. Schematic representation of the most probable electron density distributions for various types of bonds within a hypothetical two-dimensional lattice. (1) Ionic bonding. (b) Covalent bonding. (c) Delocalized bonding. (d) Metallic bonding. In solids with covalent bonding the electron distributions are concentrated along bond directions as in (b), but the individual electrons are in fact delocalized; the distinction between metals and covalent solids will become clearer in Chapter 6.

surrounded itself with four neighbours (Fig. 2.1(b)), and each pair of atoms shared two electrons, then each atom would achieve a stable state. The favourable energy situation produced by a pairing of electrons in a covalent bond provides the source for the attractive forces. The repulsive forces can again be attributed to inner-core repulsions.

2.1.3. Non-localized bonding

Previously we have pictured bonding as mainly involving a redistribution of electrons between two atoms. The final electron distribution is around particular ions or between pairs of atoms. However, certain molecules, such as benzene and anthracene, contain groupings of atoms in which a stable electronic structure may only be obtained by sharing electrons, not simply between nearest neighbours, but equally amongst a number of atoms in the molecule. This results in *delocalized orbitals* containing many electrons rather than several electron-pair bonds (Fig. 2.1(c)). The bonding in metals is similar to this, except that the valence electrons are distributed equally amongst all the constituent atoms. The bond is then known as a metallic bond (Fig. 2.1(d)).

2.1.4. *Van der Waals bonding*

We commented earlier on the chemical stability of the inert gases. However, at low enough temperatures even these elements can be induced to form condensed states. Whereas we can account for repulsive forces between such atoms by repulsions between filled atomic orbitals, the source of attractive forces is less obvious. Owing to the motion of the electrons about a nucleus in an isolated atom, there is a change with time in the density of the electron cloud. This may, at any instant, constitute a separation of charges (called a *dipole*) which can interfere with the electronic motion in neighbouring atoms. The resulting attractive force binds the atoms together.

Obviously, this behaviour is not restricted to the inert elements. A molecule which possesses a permanent charge displacement may itself induce oscillating dipoles in otherwise non-polar molecules or even interact with other dipolar molecules. The collection of all these dipolar forces constitute Van der Waals bonding.

2.1.5. *Hydrogen bonding*

A peculiar type of bond, which is believed to be largely electrostatic in origin, may be formed by hydrogen atoms under special circumstances. The electronic structure of hydrogen suggests that its reactions should be aimed at gaining or sharing one electron. Nevertheless, hydrogen is capable of forming a *bridge* or bond between two molecules. A good example is the binding of H_2O molecules in water or ice. Oxygen attracts electrons more readily than hydrogen and each water molecule acquires a permanent electrical dipole. These dipoles interact with neighbouring dipoles forming a hydrogen bridge. A hydrogen bridge can only be formed between molecules containing light elements like F, O, or N, which have a high affinity for electrons. Although stronger than Van der Waals bonds the hydrogen bond is still weak; yet it is very important in nature. Without such bonding water would boil well below $0°C$ and life as we know it could not exist.

2.1.6. *Bonds in real matter*

Some indication of how atomic orbitals are perturbed on bonding is given above. However, we have said little about the degree of perturbation involved. Now, by raising the temperature of condensed matter, we introduce energy into the system. This tends to weaken and ultimately disrupt chemical bonds in the material, turning it into vapour. The temperatures at which materials boil (or sublime) give some measure of the degree of perturbation (i.e. relative *bonding energies* or bond strengths) between the particles of the materials.

In Table 2.1 we have listed boiling temperatures for various substances, indicating the relative strengths of different types of chemical binding. The presentation of different types of bond should not be taken to suggest that each material possesses only one type of bond. In practice, combinations of

TABLE 2.1

At their boiling (or sublimation) points, materials change from the condensed to the vapour state. The change in atomic order involves the breaking of bonds. Melting temperatures also give some indication of bond strengths but do not indicate bond disruption.

ELEMENT OR COMPOUND	PREDOMINANT TYPE OF BONDS BROKEN ON BOILING	MELTING TEMPERA- TURE (Kelvin)	BOILING TEMPERA- TURE (Kelvin)
Carbon (diamond)	covalent	3773	4173
Germanium	covalent	1685	2775
Silicon	covalent	1231	3123
Potassium Fluoride	ionic	1120	1775
Sodium Chloride	ionic	1073	1738
Sodium Nitrate	ionic	581	(decomposes at 653)
Iron	metallic	1808	3073
Gold	metallic	1336	2933
Sodium	metallic	371	1156
Potassium	metallic	335	1033
Water	hydrogen bonds	273	373
Hydrogen Fluoride	hydrogen bonds	180	293
Krypton	Van der Waals	116	120
Argon	Van der Waals	84	87
Neon	Van der Waals	24	27

different types of bond occur. In water, for example, the covalent O—H bond involves a sharing of electrons between oxygen and hydrogen, although the electron cloud is displaced towards the oxygen atom. As we have already mentioned, however, it is hydrogen bonding that holds the water molecules together, and it is the hydrogen bonds that are broken when water boils.

2.2. Atomic order

Suppose that by a magical act we may isolate a medium from all external sources of energy. The material will lose energy until it reaches a certain minimum energy condition in which the arrangement of atoms is such that all attractive and repulsive forces are balanced. If we introduce energy into the system (for example, by raising its temperature) then the extra internal energy must be accommodated by a changed configuration of atoms. The number of possible ways in which the structure may be distorted in this higher energy state is a measure of the disorder introduced into the ideal structure by the intake of energy. The degree of disorder present is measured in terms of the multiplicity of available structures, and the yardstick is dubbed *entropy*.

In the following sections we shall attempt to classify materials in terms of their degree and type of order. It is convenient to think of three broad classes: crystalline, quasi-crystalline, and non-crystalline.

2.2.1. *Crystalline materials*

Perfect order would result in the ideal crystal. It is, however, the theorist's dream. In a perfect crystal, one visualizes a material of infinite extent which is completely screened from external energy sources. The high degree of order leads to a structure which is regular and *periodic*. It is possible to characterize the crystal through a volume of space, known as the *unit cell*, such that it contains sufficient atoms to specify completely the relative positions of the constituent atoms, as dictated by the types of bonding.† The unit cell, when repeated in three dimensions, is thus able to generate the complete crystal (Fig. 2.2). This ideal structure provides the basis for models of *real* crystals. Because of the high degree of order, the ideal crystal is taken as a definition of the state of zero entropy.

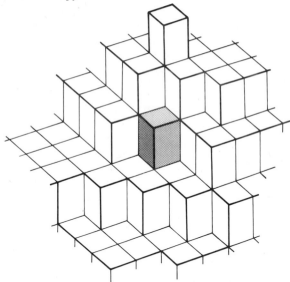

FIG. 2.2. The formation of the ideal crystal lattice by repeating the unit cell (shown shaded) in three dimensions.

Whereas both real and ideal crystals depend upon their atomic composition, the structure of real crystals is additionally affected by their method of formation, their subsequent treatment, and the final external conditions. These

† The arrangement of atoms within a unit cell has been used to subdivide ideal crystals into several crystal classes.

last mentioned factors lead to imperfections in the ideal lattice which in turn can greatly influence the physical appearance and properties of the substance. Hence, in order to understand the properties of real crystals, we need to consider the types of defects which may occur. In later chapters, we shall indicate how these imperfections affect the properties of a material, and how certain of them can be introduced into substances to produce required physical properties.

Lattice vibrations. By definition, the 'ideal crystal' is a structure with zero entropy. This means that it is imagined to be at the absolute zero of temperature. In all crystals, including the ideal crystal, the atoms vibrate about their equilibrium positions. A real crystal at a non-zero temperature must have gained additional energy. The mechanism for the storage of much of this energy is provided by an increased vibrational energy of the atoms of the crystal. The greater oscillations of the atoms constitutes greater disorder in the crystalline array and may thus be considered as a defect in the structure.

The vibrational energy is measured in terms of a quantity called the *lattice specific heat*. In much the same way as the oscillatory motion of individual water molecules can lead to waves which flow across a pond, so the individual atomic motions in a crystal are viewed as *lattice waves* which travel through the material.

Point defects. A point defect is a disturbance of the lattice structure at a single point in the crystal. There are essentially three types of defect which may occur (Fig. 2.3). Firstly, a foreign atom may have replaced one of the lattice atoms. We then speak of *impurities* or *substitutional defects*. As we shall see in Chapters 4 and 5, minute quantities of certain such defects may profoundly alter particular properties of crystalline materials. Secondly, an atom (or ion) may be lodged in an *interstitial* position between normal lattice sites. The vagrant atom may be a displaced host atom (or ion) or a chemically different intruder. Thirdly, we have possibly the most obvious form of defect: namely the absence of a host atom (or ion) from its appointed position, aptly termed a *vacancy*.

In accommodating the vacancy, the crystal may create an extra atomic site at its surface. This is then called a Shottky defect. Alternatively, the displaced atom may go to an interstitial position whence we have a Frenkel defect. In an ionic crystal, the vacant site has an effective charge relative to that of the perfect lattice. The lattice resents this and endeavours to remain neutral: for Shottky disorder there are therefore equal numbers of positive and negative ion vacancies in the absence of other impurities.

Line defects. Line defects are essentially two-dimensional crystal defects. We shall only attempt to give an elementary illustration of this type of disturbance. Consider a plane ABCD within the crystal and let an area $e_1s_1e_2s_2$ lie within this plane as shown in Fig. 2.4(a). By divine action let us displace the

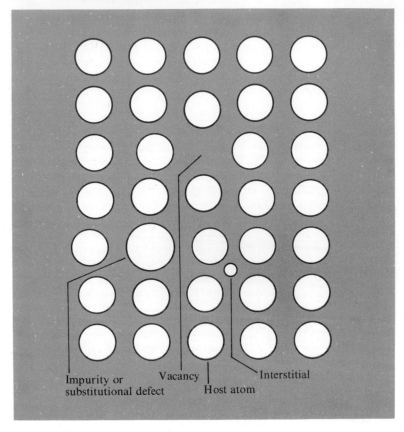

FIG. 2.3. Point defects in a hypothetical two-dimensional crystal lattice.

material located within, and above, the area $e_1s_1e_2s_2$, relative to the rest of the crystal, by an amount b, in the direction indicated by the arrow in Fig. 2.4(b). This creates regions of compression and extension within the crystal. To neutralize these, the lattice atoms move as indicated in Fig. 2.4(c). One can see that the greatest degree of distortion occurs around the boundary of the area $e_1s_1e_2s_2$. Such boundaries throughout the crystal are termed dislocations. Early workers used the Italian word *distorsioni*, which better describes the situation.

In general the distorted atomic arrangement around the dislocation line is complex. However, it is sometimes possible to resolve this complex structure into one or other, (or a combination of two) comparatively simple types of distortion. These simple examples occur when the direction of displacement

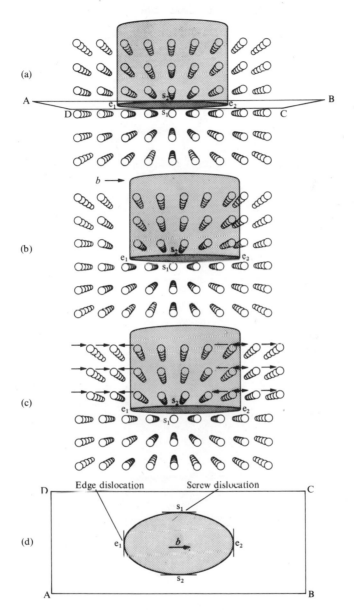

FIG. 2.4. Dislocation lines in a three-dimensional crystal lattice. (a) The undeformed crystal lattice. (b) Regions of compression and extension are introduced by shifting certain atoms from their equilibrium positions. (c) Lattice atoms move in an attempt to repair the introduced distortion. (d) Plan view of the dislocation line showing the separate edge and screw components.

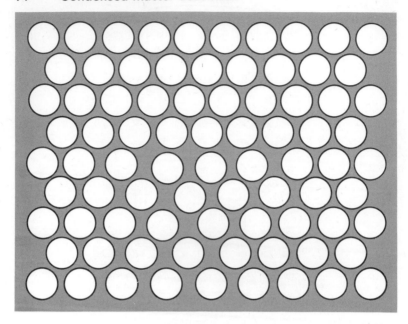

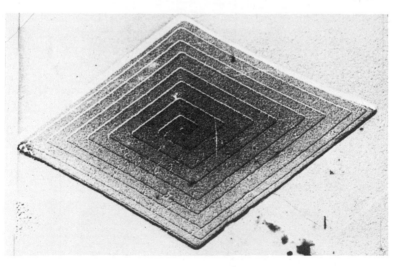

FIG. 2.5. Edge and screw dislocations. (a) Schematic diagram of an edge dislocation as it might be seen in a section through a metallic lattice. Tilt the page and view along the lattice lines. (b) Spiral growth from a screw dislocation in a single crystal of n-paraffin (Courtesy of A. J. Forty).

or slip, signified by the arrow in Fig. 2.4(b), is parallel or perpendicular to the dislocation line. In the former case we speak of a *screw dislocation*, and in the latter case of an *edge dislocation*. Thus in Fig. 2.4(d) there are screw dislocations at points s_1 and s_2 and edge dislocations at points e_1 and e_2. Between these various points, and around the dislocation line, there are mixtures of edge and screw dislocations.

Both edge and screw dislocations are comparatively easy to picture on their own. An edge dislocation arises apparently from the presence or absence of a plane of lattice atoms (Fig. 2.5(a)). The screw dislocation is best visualized where it meets a surface. Here it provides a site upon which the spiral growth of a crystal from the vapour phase may readily occur (Fig. 2.5(b)).

Surfaces and interfaces. An obvious distinction between a real and the ideal crystal is the presence of a surface. The outer extremities of the real crystal provide an abrupt termination to the essentially periodic structure. One may also encounter, *within* the material, junctions or areas across which there is a marked change in the nature of the crystal. Such *interfaces* may occur naturally, or, as we shall see in Chapter 5, may be created artificially. Natural interfaces usually involve changes in the orientation of the lattice planes of atoms. The complete crystal may then be a mosaic of many such small crystallites or crystal grains as they are often called (Fig. 2.6). In such a case, the structure is termed *polycrystalline* and the interfaces are called *grain boundaries*.

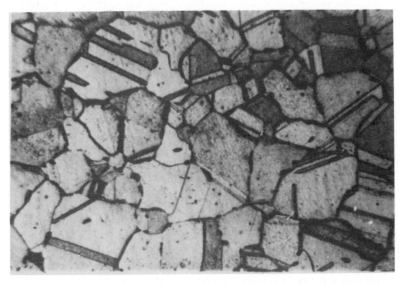

FIG. 2.6. Grains in alpha-brass, magnified about 150 times. (Courtesy of P. Feltham.)

Subatomic defects. External energy may also be absorbed in a crystal by exciting certain constituent electrons to higher energy states. The ability to store energy in this way is measured in terms of an *electronic specific heat*. In addition, short-lived quantities of energy may be present in the lattice in the guise of photons (packets of electromagnetic energy), excess electrons, or neutrons following the bombardment of the lattice with suitable radiations. Such defects owe their origin to atomic (rather than crystalline) properties. They are all short-lived effects.

2.2.2. Non-crystalline materials

Perfect order and perfect disorder are embodied in the ideal crystal and the ideal gas respectively. Both of these limiting cases provide models for real systems. Non-crystalline, or *amorphous* materials as they are generally termed, fall between these two extreme models. With amorphous materials there is no real evidence of a periodic structure. Nevertheless, there is evidence for the preferential ordering of neighbouring atoms around a particular atom or ion. One talks in terms of a preservation of *short-range order* but a loss of *long-range order* throughout the material. The two most common classes of amorphous materials are liquids and glasses.

Liquids. In a liquid, the atoms are sufficiently close together for interatomic forces to influence their relative spatial array. However, in this class of material the internal energy stored in the medium is comparable with the interatomic bonding energies. Because of this, no extensive networks are formed. Thus short-range, but not long-range, order exists. The nature of the short-range order in liquids is the topic of current research. The two commonest models have been developed from models for gases on one hand and crystals on the other. The quasi-crystalline approach views the liquid as a degenerate crystal in which the density of defects has risen to such an extent that the long-range periodic structure has been obliterated (Fig. 2.7(b)) whilst short-range order is retained in local regions. The gas-like model starts with a random arrangement of atoms and considers the interatomic forces to cause some local bunching or grouping of the atoms (Fig. 2.7(c)). Once again, each model has its advantages and disadvantages, and the true picture lies somewhere between the two.

Hitherto, we have described the liquid state simply in terms of the atomic order. Some physicists feel this to be inadequate. Whereas they believe that an instantaneous picture of the atomic array would satisfy the foregoing description, the structures themselves have only a transient existence. They suppose that there is a constant change in the atomic positions whilst the same average number of atoms is maintained in a local region. Observation of a finite volume in the fluid would show the continuous formation and destruction of such groupings. This type of model seems most appropriate to liquids in which the maximum packing or coordination is restricted mainly by repulsive

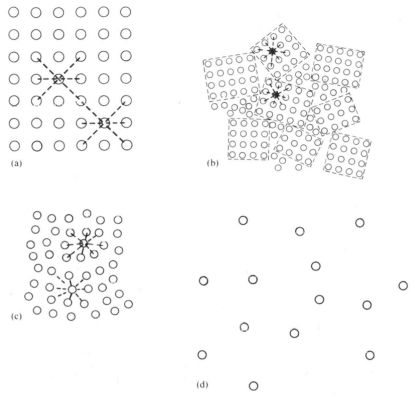

FIG. 2.7. Two-dimensional representations of models of the liquid state. These stem from the familiar pictures of (a) the ordered crystal, and (d) the disordered gas. The quasi-crystalline model (b) pictures the liquid as a highly disordered solid whereas the gas-like model (c) considers the liquid as a partially ordered gas.

forces. Examples include liquid metals and liquid inert gases. Other physicists visualize liquids as consisting of specific regions within which the local order is well defined and maintained with time. The random distribution of such units accounts for the loss of long-range order. This approach is better suited to explaining the properties of liquids in which attractive forces appear to dominate the short-range order. Examples include hydrogen-bonded H_2O molecules in water at low temperatures, the ring structure of molten sulphur, and the existence of chains and rings in liquid selenium. Certainly there is evidence for a greater degree of order in these liquids. In fact the liquid state, in its many forms, probably spans the gap between a very highly disordered solid and a moderately ordered gas.

Glasses. Glassy or vitreous materials are very familiar in our everyday experience. They are produced by rapidly cooling (i.e. quenching) certain types of liquid. The resulting solid has a liquid-like structure in that it contains only short-range atomic order, but it is formed at such a temperature that the internal energy is not sufficient to compete with the atomic bonding energies. A cursory thought might lead us to suspect the existence of the vitreous state. For example, why don't the atoms assume their appropriate crystalline positions on cooling? The answer lies in the method of formation of the glass. If we cool a liquid, we drain away some of its energy and thereby restrict the motion of the atoms. Slow cooling allows sufficient time for the atoms to move to positions consistent with the temperature of the melt. In contrast, rapid cooling catches the atoms unawares, so that they are unable to move to their crystalline positions before they have lost the necessary energy. A glass is thus a 'frozen' liquid, in which atomic motion has been greatly slowed down. It is interesting to note that, if left for a sufficiently long time, the glass will gradually adopt the crystalline state. Hence the glassy state is strictly a *metastable* rather than a stable one.

2.2.3. *Quasi-crystalline materials*

This type of material refuses to fit neatly into either the crystalline or amorphous class. It embraces materials in which certain groups of atoms show a crystal-like structure whilst other regions appear amorphous. Typical of such structures are polymers, liquid crystals, and super-ionic solids.

Polymers. The essential unit of a polymer is a chain of atoms in which a certain basic structure is repeated a large number of times. A one-dimensional equivalent of an ideal crystal is a good picture of the chain. However, despite the simplicity of this chain, the combination of many such chains in a confined volume, produces a complex and varied structure. The bonding within a single chain is usually covalent and strong. That between chains is generally weak, being either of the Van der Waals type, or even simple electrostatic attraction between charged groups. The interchain attraction clearly depends on the chain length.

For any assembly of small chain molecules, the interchain forces are generally too weak to cause chain alignment (Fig. 2.8(a)), but as the chain length increases, the inter-chain binding and hence the molecular alignment increases (Fig. 2.8(b)) and the structure shows increasing crystallinity. In the case of polymers, which are composed of long chain molecules, we might think that the very long chains would lead to a high crystallinity. However, just as a short length of thin wire a few centimetres long appears rigid whilst a metre length is very floppy, so the polymer chains become flexible with increasing length. The resulting coiling of the chains destroys the regularity of the medium (Fig. 2.8(c)) and gives rise to local amorphous regions. Clearly, the properties

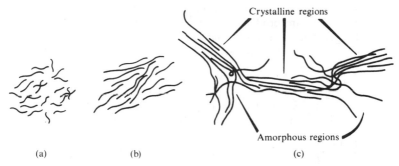

Fig. 2.8. Variation in structure of a polymer with increasing chain length. (a) Randomly oriented short chains. (b) Increased chain length increases chain alignment. (c) Twisting and coiling of very long chains prevents complete alignment.

of solid polymeric materials depend on much more than their chemical composition. The skill of the polymer chemist involves choosing types of polymer chains and attempting to vary the intra- and inter-chain bonding to suit his needs. For example, freely coiled chains permit the material to be extended under tension, and spring back when released. Such is the behaviour of *rubbers*. Other materials exist with their chains held in the highly extended state, often as a result of inter-chain forces. The improved order in these materials is accompanied by strength in the direction of the chains. Natural examples of such *fibrous* structures are wood, cotton, silk and muscle tendon. Synthetic polymers, such as nylon and polyesters, are often drawn into fibres during the manufacturing process. Materials with properties intermediate between those of rubbers and fibres can generally be moulded easily into various shapes and are called *plastics*. Examples are polythene, polyvinyl chloride and polystyrene.

Cross-linked polymers can be made by using suitable chemical starting materials. These have a three-dimensional structure which cannot be melted by heating. An example is epoxy resin adhesive.

The term polymer thus indicates a wide and varied collection of materials. In a text of this size we can only try to indicate some of the properties of this inexhaustible material class.

Liquid crystals. The apparently self-contradictory name 'liquid crystal' is used to describe a group of materials which can be poured from a container like a liquid and yet consist structurally of a loose array of molecules which is in many ways similar to a crystal lattice. In this case, the 'crystal' is formed by whole giant molecules, rather than atoms. The materials are composed of large rod-like or ellipsoidal molecules. These are well aligned and ordered in a rather close-packed manner. However, the molecules are permitted to move within the body of the material. It is the degree of freedom of the

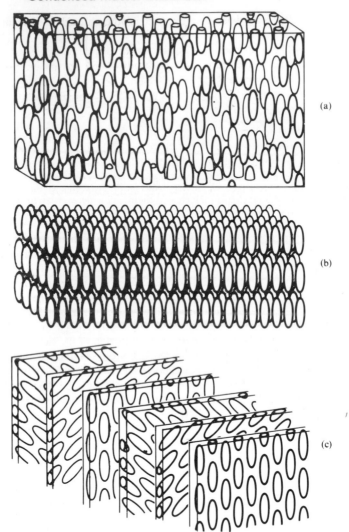

(a)

(b)

(c)

FIG. 2.9. Structure of liquid crystals. Relative arrangements of molecules in (a) nematic, (b) smectic, and (c) cholesteric forms.

molecules which differentiates between the major liquid crystal classes. If the molecules can move both up and down and sideways, as in Fig. 2.9(a), the material is *nematic*. If the vertical motion is severely restricted, we obtain a layered (*smectic*) material (Fig. 2.9(b)). The third class is the *cholesteric* in which parallel layers are twisted relative to each other throughout the material (Fig. 2.9(c)).

The crystal-like properties are sensitive to the degree of molecular order and hence to the temperature. This is demonstrated by the increasingly utilized property of certain thin films of cholesteric liquid crystals to change colour with variation of the temperature of the substrate on which they are formed.

Super-ionic solids. Certain materials like Ag_3SI and $NaAl_{11}O_{17}$ also possess properties somewhat intermediate between solids and liquids. In $NaAl_{11}O_{17}$ for example, the aluminium and oxygen atoms form a rigid lattice, characteristic of a solid. Within this framework, sodium ions are scattered in a random fashion. Under the inducement of an electric field, these ions may flow like a liquid through the solid lattice, rather as water might flow through a honeycomb. This new class of material promises to be useful in rechargeable solid-state batteries, and in miniature, high-storage capacitors.

2.3. Changes in atomic order

From Chapter 1, we recall that a given material may be encountered in one or more of the solid, liquid, or gaseous states. The natural state of atomic order for a collection of atoms depends on the internal forces and the external conditions. In the absence of external energy sources, internal forces determine the structure. By considering different possible homogeneous atomic arrangements, or phases as they are called,† we may calculate the probable total energy and hence the average atomic separation for these structures. The structure which corresponds to the minimum energy will be the one finally adopted. In Fig. 2.10(a), various phases are assumed to correspond to positions I to IV on the interparticle energy diagram. Under the conditions of the diagram, the structure would exist in state II. Let us then raise the temperature. The energy supplied from the exterior will be accommodated as increased thermal vibrations of the atoms. These thermal vibrations correspond to what are called 'excited states' of the structure II (Fig. 2.10(b)). In Fig. 2.10(b), we have also pictured hypothetical excited states for the possible phases III and IV at various temperatures (T_1, T_2, T_3 and T_4). At temperatures T_1 and T_2, the energy of the excited states in phase II are still lower than the excited states of phases III and IV. Hence phase II is still energetically the most favourable. However, at temperature T_4, phase III is at a lower energy than phases II or IV. Thus at the temperature T_3, where the excited states in phases II and III are equal, the system changes its structure. Such a transition is called a *phase transition*. Let us review the mode of this change. At a temperature T_1 the material exists in phase II. Heating the material raises the

† Strictly, a phase is defined as a part of a system which is homogeneous, physically distinct, and isolated by a definite boundary.

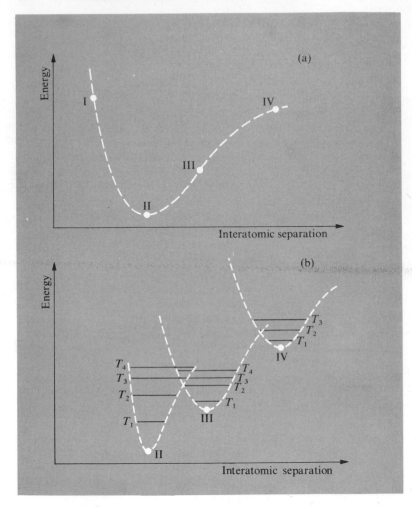

FIG. 2.10. Idealized phase equilibria in materials. Schematic plots of energy against interatomic separation for hypothetical atomic arrangements (a) at absolute zero, (b) at finite temperatures. This second graph shows the excited states present at various finite temperatures T_1, T_2, T_3, and T_4 with $T_4 > T_3 > T_2 > T_1$.

temperature of phase II to T_3. At this temperature, further heating provides energy for the atoms to reshuffle into the higher-energy positions characteristic of phase III. When the change is complete, further injection of heat simply raises the temperature of this material in phase III.

The process is possibly more readily visualized if we consider a system containing H_2O molecules and call phase II ice and phase III water. In addition to temperature, pressure variations affect phase changes. An example of this is the local melting of ice under the blade of a skater's boot. Such a local phase change in the ice enables one to skate on this 'solid'.

The extra energy liberated or absorbed as heat during the transition between phases II and III at a fixed temperature T_3 is called a *latent heat*. This energy is used to change the particular bonding network. Hence, the temperature (at a fixed pressure) of a phase change is an approximate indication of the type of bonding involved (Table 2.1).

So far, we have considered only the normal or equilibrium states of systems. How do we account for the existence of such metastable states as the glassy state? Let us imagine phase III (Fig. 2.10(b)) to be the liquid state and phase II to be the crystalline state of a glass-forming material. By quenching the liquid suddenly from temperature T_4 to T_1, we prevent the system from transforming at temperature T_3 and so freeze its structure in the unfavourable excited state represented by T_1 in phase III. The glassy state is thus seen to be a 'liquid structure' which due to its reduced internal energy possesses a very high viscosity and many 'solid-like' properties.

3. Elastic properties of materials

'And so no force, however great, can stretch a cord, however fine, into a horizontal line which shall be absolutely straight.'

WILLIAM WHEWELL *Elementary treatise on mechanics*

MAN has always been interested in the mechanical properties of materials. In early history he searched for 'hard' materials for making tools, 'strong' materials for buildings and fortifications, and free-flowing substances for moulding and pouring. In this chapter, a limited discussion is given of the origins of these various properties in real materials. This leads to the consideration of new materials with improved mechanical properties. In general, both the bulk and the surface properties may be traced back directly or indirectly to the forces between the constituent atoms of the materials.

Forces can be applied to a body in a number of ways. An unopposed force will cause a body to move. However, two or more opposing forces may leave the body stationary but subject it to a stress (Fig. 3.1). One may consider that

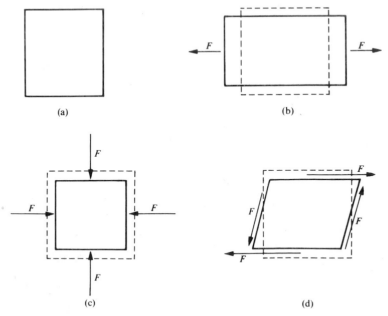

FIG. 3.1. Equal and opposite forces (F) applied to a body. (a) Unstressed body with $F = 0$, (b) tensile stress, (c) hydrostatic stress, and (d) shear stress in the body.

materials respond to an applied stress in one of three ways. Firstly, if the stress is not too great, the material deforms and reaches an equilibrium situation which is maintained until the stress is removed. The material then reverts to its original shape and state. Under these conditions, the material is said to be *elastic*. If the material does not return to the original state upon removal of the stress, but assumes some degree of permanent deformation, it is said to be *plastic*. Finally, the sample may never reach a static equilibrium, but may continuously deform under the influence of the applied stress. The material then flows and gives rise to *viscosity* effects.

The magnitude of the stress required to produce these effects varies with the material and its structure. As mentioned in Chapter 1, systems which tend to flow are those with reduced atomic order due to a relatively high internal energy. High crystallinity, however, is indicative of well ordered, closely packed atoms which are relatively difficult to displace. One therefore expects predominantly to encounter viscosity effects in liquids and elastic properties in solids.

3.1. Elasticity

3.1.1. *Ideal crystals*

When applied to a crystalline solid, tensile (or compressive) stresses tend to alter the separation between pairs of atoms along the lines of action of the opposing forces. Throughout the length of the material this results in an extension (or contraction) of its dimension. Variation of the interparticle separation results in a change of the bond energy from its minimal value (Fig. 3.2) with the result that restoring forces are brought into play. These will

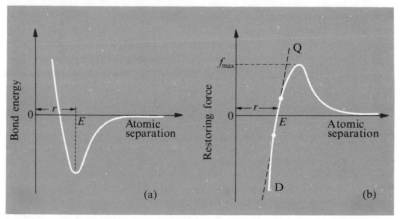

FIG. 3.2. Localized bond energy (curve a) and restoring force (curve b) as a function of the interatomic distance in an ideal crystal. The restoring force is zero when the energy is a minimum.

be attractive if the deforming forces are tensile or repulsive if they are compressive. An equilibrium condition is maintained between the deforming and the restoring forces as long as the applied stress is maintained. Upon removal of the applied forces, the interatomic separation returns to the value corresponding to the minimum of the energy well of Fig. 3.2 under the influence of the restoring forces. As this happens to all atomic pairs throughout the crystal, the material contracts back to its original length. In the region of the field-free equilibrium position (E of Fig. 3.2), the restoring force is approximately proportional to the local deformation. For the complete crystal, the overall deformation is proportional to the total restoring forces and hence to the deforming forces. Hence, for an ideal crystal, the behaviour of the interatomic forces is consistent with Hooke's law, which states that for small applied forces, the elastic displacement is proportional to the magnitude of the applied forces. From Fig. 3.2 we note two further interesting points. Firstly, if the applied forces are increased, corresponding atomic restoring forces will ultimately leave the linear (QED) section of the force curve. The behaviour may still be elastic, but with a non-proportional relationship between the applied forces and the deformation. Secondly, indefinite increase of the tensile forces cannot be sustained. In extension, there is a maximum force, f_{max}. If this condition could be reached, the atoms would separate and the material would disintegrate. In practice the substance breaks at a weak point before this situation is obtained.

3.1.2. *Elastic moduli*

When a solid rod of length L and uniform cross-sectional area A extends by an amount l under tensile forces F, then the tensile stress applied to the rod is F/A and the tensile strain is l/L. The quotient of this stress to this strain is called the Young modulus (Y) of the material of the rod, provided that F is not so large as to exceed the conditions of Hooke's law. From the foregoing section and Fig. 3.2(b), we see that Y is related to the nearly linear response of the restoring forces with distance in the neighbourhood of equilibrium. So far, we have only discussed the effect in the direction of the applied stress. In a crystal lattice, atomic displacement along the line of action of the opposing forces must have an effect on lateral bonds. In fact, as the body extends longitudinally under tension, it contracts laterally. The fractional decrease in the lateral dimension is the contractile strain, and the ratio of contractile strain to tensile strain is known as the Poisson ratio. For most elastic materials, it varies from 0·2 to 0·4. A value of 0·5 implies that there is no loss of volume upon extension or compression. This situation is almost realized with rubbery polymers.

Two other moduli are in common use. The first is the bulk modulus, for which the forces are applied to all faces simultaneously. Hence, there is a hydrostatic pressure (or stress) on the material. The bulk strain is then the fractional volume change of the material. The reciprocal of the bulk modulus is called the compressibility of the material. The second modulus is the rigidity modulus, which occurs when forces F are tangentially applied to opposite faces of area A of the material so as to constitute a shear stress of F/A. The angle of shear is taken as the strain. The ratio of the appropriate stresses to strains defines the respective moduli.

3.1.3. *Elasticity of solids*

From the ideal crystal, one would expect that crystalline materials with strong atomic bonding would have strong restoring forces for a given inter-atomic displacement of their atoms from their stress-free positions. Hence, from the definition of the Young modulus, they might have a high value of Y. In many common materials, the bonding is not exclusively of a single type. However, the melting temperature (T_m) is a useful indicator of the strength of the bonding in a given substance. In Table 3.1 it is seen that as a general rule, the higher the melting temperature, the greater is the Young modulus. It is interesting to note that the metallic bond is relatively strong,‡ as is reflected in the values of Y. For comparison, the table includes some complicated but

TABLE 3.1.
*The Young modulus (Y) and the melting temperature (T_m)
for various materials.*

MATERIAL	T_m (kelvin)	Y $(10^{10} \text{ N m}^{-2})$
Diamond	3773	84
Carbon whiskers	3773	77
Boron	2300	41
Sapphite (aluminium oxide)	2307	40
Silicon	1685	16
Titanium	1941	12
Copper	1356	13
Glass (crown)	†	7·1
Aluminium	933	7·0
Magnesium	923	4·5
Wood (pine) {along grain	†	1·6 }
{radially	†	0·11 }
Sodium	371	0·9
Polystyrene	†	0·4
Rubber (natural)	†	0·0007

† These materials do not have a melting temperature; see text.

‡ In Table 2.1 we saw that the boiling temperature was more appropriate than T_m in the case of metals.

common materials. It should be noted that glasses and polymers do not melt; they soften. A glass is already a 'liquid' (section 2.2.2) and an increase of temperature predominantly affects the sample viscosity. Polymers consist of both crystalline and amorphous regions (section 2.2.3). Whereas the crystalline regions melt at a definite temperature, the amorphous regions 'soften'.

Many solids have an anisotropic structure in which different bonds, or the density of various bonds, vary with crystal directions. It is not suprising to find therefore that the elastic modulus is also anisotropic for such solids (see

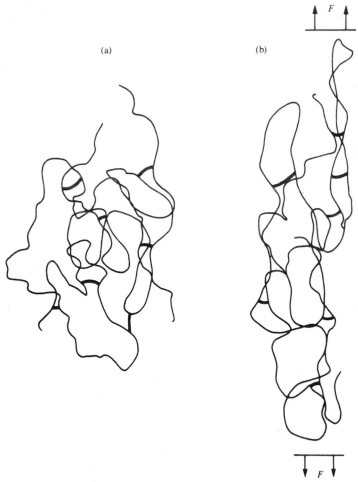

FIG. 3.3. Schematic diagram of cross-linked polymer molecules in the rubbery state; (a) and (b) represent the absence and presence of extending forces respectively.

Table 3.1). With graphite, Y varies by a factor of some thirty times for different crystal axes. If the individual grains of polycrystals have an ordered array, Y is anisotropic, whilst randomly oriented grains result in an isotropic modulus.

The elastic properties of rubbers are particularly interesting owing to the ability of the polymer chains to deform and glide over each other (Fig. 3.3). This property accounts for two of the remarkable elastic characteristics of rubbery polymers. Firstly, Y is rather small and the rubbers have a remarkable extensibility. It is not uncommon to stretch a rubber up to three or four times its original length, and still remain within the elastic limit. A few per cent increase in length is all one might expect for the other materials listed in the table. Secondly, the stretching of the polymer chains imposes order on the system. High temperatures are the enemy of order. Hence, increasing the temperature opposes the action of the applied tensile stress and the Young modulus *increases* with temperature. Rubbery polymers are the only class of material that have this characteristic.

3.1.4. *Elasticity of liquids*

Because of their tendency to flow, liquids cannot be made to reach an equilibrium state under a shear stress. Similarly, it is generally unlikely for them to support a tensile stress. They can, however, be subjected to hydrostatic pressures, and their bulk modulus or compressibility determined. In all cases, the bulk modulus is very high (or the compressibility very low) when compared with a gas. It is important to note that, in the liquid state, the compressibility is dependent upon the conditions being isothermal or adiabatic.†
The relative incompressibility of liquids has a number of applications in engineering.

3.2. Plasticity

3.2.1. *Plastic deformation*

The interatomic force diagram (Fig. 3.2) indicates that, in ideal crystals, a somewhat catastrophic fracture might occur if external forces are increased to the extent where Hooke's law is exceeded and the retaliatory restoring forces approach a total value of f_{max}. In practice, this does not usually happen with metals and polymers. Solid metals, and to a lesser extent crystalline polymers, have a highly regular and organized atomic structure. Yet both of these materials have the important property that they can be extended greatly beyond the proportionality limit (the limiting condition up to which Hooke's law is obeyed) to a region in which the material responds to the applied stress by permanently setting. This is well before the material fractures. In this region, the material is said to undergo *plastic deformation* and it is this property which allows metals and plastics to be rolled, extruded, moulded, and beaten

† Isothermal conditions exist when heat is allowed to enter (or leave) the system and maintain the temperature constant. Adiabatic conditions imply the prevention of heat exchange with the surroundings.

into shape without the materials springing back to their original dimensions. Does this behaviour signify the collapse of the interparticle-force hypothesis? Is the situation represented in Fig. 3.2 completely invalidated?

Before considering these questions, let us consider the behaviour of a metal. An illustrative stress versus strain curve is shown in Fig. 3.4. Plastic deformation takes place in the region P to Z where the stress required to increase the deformation increases with the deformation. This is termed *work hardening*. The metal will not stand unlimited stress and at Z, it suddenly *thins* very abruptly. This is called *necking* and fracture is imminent. If the stress is removed at some point R in the plastic region, the sample contracts back to a condition such as O′. The path O′R is approximately parallel to OL and represents partial elastic recovery. The sample has permanent deformation corresponding to OO′ which is not recoverable by further loading and unloading. Upon reloading, the path O′R is approximately retraced (except near R) and the original work-hardening curve is rejoined. Before leaving this curve, note the break in the abscissa of the graph. A metal specimen may extend by a few per cent in the elastic region, but by many times its length in the plastic region.

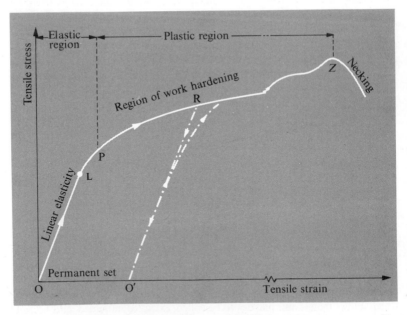

FIG. 3.4. Typical stress–strain curve for a metal. OL represents the linear elastic region. At P, the yield point, plastic deformation occurs and continues by 'work hardening' up to Z.

3.2.2. *Origins of plasticity*

There are at least three clues to the origin of plastic deformation. Firstly, theoretical estimates predict a yield stress (as at P in Fig. 3.4) for crystals which is some 10^3 times the observed values for bulk metals. Secondly, very pure metal 'whiskers' have elastic properties much closer to the predicted limit than bulk metals. Hence, it would seem that the interatomic force hypothesis can apply, but is not the only factor with real crystals. Thirdly, we note the surface morphology of certain materials after plastic deformation (Fig. 3.5).

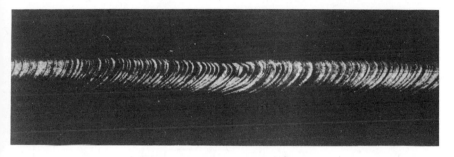

FIG. 3.5. A 'slip-deformed' single crystal. Note the steps produced on the surface. (Reproduced from C. F. Elam (1935) *Distortion of metal crystals*, Clarendon Press.)

The major origin of deformation is the ability of certain regions or blocks of the crystal to slip past each other once a certain tensile stress has been reached. In each block, the atomic structure is preserved, although the atoms suffer the elastic deformation experienced before the onset of the yield point. Such blocks are termed *glide packets*† and are typically a few μm (i.e. 10^{-6} m) in width. The common surface over which the slip occurs is the *glide surface* or, when it is a discrete plane, as is often the case, it is termed the *glide plane*. Sometimes the packets are separated not by a single glide plane, but by a *zone* of a few close crystallographic planes.

When glide occurs the atoms on one side of the plane shift an integral number of atomic spacings, so that the localized, microscopic structure is completely restored. What happens is that the surface of the sample becomes staggered by a whole number of atomic displacements at each step. From a crystallographic viewpoint, slip could be initiated on any plane in the material. In practice, for a given material, the glide planes are in a particular crystallographic direction, which is usually that with the greatest density of atoms. It should be noted that the glide surface is not along the direction of the applied stress, but at an angle to it. Hence the actual stress experienced by the glide

† The terms 'slip' and 'glide' are equivalent, and both are used in the literature.

surface is a *shear stress* and not a tensile one. It is because the glide is at an angle that the material becomes extended in length (Fig. 3.6). Because the localized atomic order is complete after the glide, the *status quo* is maintained with removal of the applied stress and the plastic deformation is not reversible. The restoring forces relax, however, and the elastic deformation is reversible.

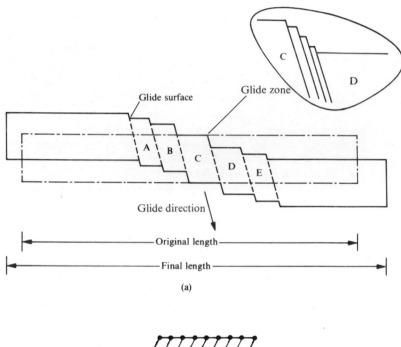

(a)

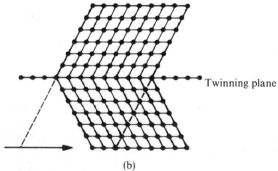

(b)

FIG. 3.6. Diagramatic representation of deformation mechanisms in crystals: (a) glide, and (b) twinning. In (a), A—E represent glide packets.

Certain materials deform by *twinning* rather than by glide. Under an abrupt stress, the atomic layers below a certain plane (called the twinning plane) move by varying amounts in the same direction, parallel to the plane. The displacements are *not* a whole atomic spacing, but are such that the displaced atomic layers below the plane adopt mirror image positions to the undeformed material above it. Twinning occurs with abrupt stresses and accounts for the creaking noise that may be heard when polycrystalline tin is bent.

Glide does not occur by the simultaneous displacement of all atoms in the glide plane. This would require the instantaneous fracture of an enormous number of bonds. In fact, glide starts at local flaws in the structure—the dislocations discussed in Chapter 2. For example, in the region of an edge dislocation, there is a localized misfit of atoms (Figs. 2.4, 2.5, 2.5(a)). If we move these atoms so as to perfectly 'fit' with their neighbours on one side of the dislocation, we simply transfer the misfit sideways. By repeating this process, the dislocation can move through the crystal and involve the disruption of only a few bonds at a time. Hence, a moving dislocation forms a boundary between the slipped and unslipped regions of the glide plane. Ultimately these dislocations would run to the edge of the material and leave a more perfect crystal. In practice, there is a mechanism whereby the dislocations multiply. We refer the reader to the bibliography for details of this process.

Increasing plastic deformation requires continued motion of the dislocations. These begin to meet and entangle, thereby restricting their ease of motion. An ever-increasing applied stress is required for plastic deformation and this is the process of work hardening. The material may be resoftened by the process of *annealing*. This consists of consecutive heating and cooling of the material. Upon heating, the system gains energy and the dislocations run back to those positions in the crystal where the lattice strain is least. Cooling 'freezes' the dislocations in these positions and the material is thereby softened.

3.2.3. *Some important terms*

The Young modulus represents the proportional amount by which a material may be extended under a given stress and yet return to its original form upon release. It is thus a direct indication of the *stiffness* of the material. The *strength* of a material is the maximum stress it can withstand before failure is encountered. In principle it is directly related to the bonding forces between the atoms. Hence, if one could obtain defect-free crystals, metals would be weaker than ionic-bonded substances, which would be weaker than covalent-bonded materials. In practice, the strength of a material is affected by the presence of defects, especially dislocations, in the crystal structure, as these allow plastic deformation to occur whilst the stress is still low. In this respect modern whisker and fibre crystals show promise as future strong materials. These are single crystals of about 1 μm in diameter which are extremely free from defects and can stand very high tensile stresses. The strength of fibres is also exhibited by amorphous and organic materials. Although they are not in the class of metal whiskers, it is interesting to note that fresh fibres of glass can stand extremely high tensile stresses when compared to the equivalent bulk materials. This is one of the properties utilized in *fibreglass* in which many fibres are bonded together laterally with resin to produce strong mats. When fibres are highly ordered, strong materials like flax are obtained, while low order, as in wool, results in materials that are weak under tension. A *brittle* material may suddenly break while still elastic, owing to the growth and propagation of cracks. Corners, flaws, or notches produce the cracks by concentrating the stresses. *Tough* materials are the opposite in that they resist crack propagation. Hence they break gradually through plastic deformation. This is illustrated by composite materials, laminates, and fibre bundles. Here the object is to stop cracks from propagating, so the bundles are laid parallel and impregnated with resin or adhesive. A crack through one bundle soon meets the adhesive interface and is arrested. A similar principle is used to *harden* (that is to increase the yield strength of) materials which undergo plastic deformation. In this case, an attempt is made to stop the dislocations from propagating by alloying the material; that is, introducing atoms which act as a solute and block the dislocation motion. Hard steel is the result of iron carbide particles in the iron lattice, whilst particles of Al_2Cu harden aluminium. Whereas these materials are hard, they are often brittle and fracture easily.

In conclusion, one can see that the plastic behaviour of crystalline materials is a direct consequence of the atomic bonding and of local imperfections in the atomic lattice.

3.3. Viscosity

3.3.1. *Liquid flow*

There are at least two views of the atomic or molecular structure of liquids. One model considers the liquid as if it consisted of random gas molecules brought into contact, but in such a way that the molecules are independent and can freely slide over their neighbours and hence through the body of the liquid. The analogy of continuously moving peas in a close-packed barrel is often used. The alternative view is that the molecules or atoms exist in small 'units' and that each unit has a lattice structure possibly typical of the equivalent solid lattice. The presence of broken bonds at the edges of the units indicate that whereas atoms might change allegiance from unit to unit, definite short-range order exists within each unit. Both models are compatible with the absence of any long-range order and the high degree of mobility in liquids, be it individual atoms or atomic units that move. It is easy to see how gravitational forces cause the liquid to adopt the shape of its container. When subjected to any shearing stress the liquid suffers continuous deformation and is unable to support the stress. Such a continuous deformation is called flow. In flow, the liquid atoms or units move past each other so that the deformation is produced by a localized shear stress. For large stresses, the flow is *turbulent* and complete mixing occurs as eddy currents are established. The atoms tumble over each other and the liquid advances in the direction of flow, with a constant velocity across its profile. Of greater interest is the consideration of small stresses. The flow is then very different, and is said to be *streamlined* or *laminar*. In this case, adjacent layers in the flow direction move with different relative velocities, the velocity being zero in the layer adjacent to the containing vessel and increasing to a maximum at the centre of flow. The interatomic forces between atoms in the adjacent layers are manifest in that the liquid opposes the flow. This opposition is an internal friction or viscous force and the material is said to possess viscosity. Under gravitational forces, water flows more readily than treacle, for example, and is thus less viscous. It is convenient to quantify this property through a coefficient of viscosity η, obtained by timing the rate of flow of the liquid through capillary tubes or from measurements of the couple opposing the motion of a cylinder which is rotated in the liquid.

Consider an atom at the centre of a group of atoms in the heart of a *stationary*, unstressed liquid. According to either the 'free-atom' or the 'lattice-unit' model of liquids, for this atom to move into a nearby vacancy, it must have enough kinetic energy to overcome the attractive forces of the neighbouring atoms. In a streamlined flow, this atomic displacement will be assisted if the applied shear stress acts in the appropriate direction. This is equivalent to saying that the energy barriers between the atomic sites are reduced in the direction of flow. Even if the appropriate vacancy is not in the direction of the shear stress, random collisions will ultimately rotate the unit (loose or bound) into a favourable orientation. The atom may then move down the flow field and into the vacancy. Meanwhile, other atoms, both in the centre and at the leading edge of the liquid, have been moving under the same mechanism. The result is a steady flow of liquid down the shear stress direction. The rate of liquid deformation is thus related to the interatomic forces and the magnitude of the shear stress.

Three further points are of interest. Firstly, the flow rate depends on the size of the vacancies. Increase of the external pressure compresses the liquid, reduces the vacancy volume, and thus increases the apparent internal friction, i.e. increases the viscosity. Secondly, increasing the temperature increases the free volume of the system. This gives rise to a greater number of vacancies so that the viscosity is reduced. Thirdly, the surface of any material must expose incomplete bonds at the outermost atoms. This will be true for the atoms constituting the wall of the container. A liquid atom is highly likely to attach itself to an unsaturated bond of the vessel wall. A layer of atoms formed from the liquid, may remain at rest on the wall during flow. Neighbouring liquid atoms will then have to drag past these stationary atoms in moving to vacant sites. Also, a cluster or unit near the wall will not be so free to turn under Brownian forces as one in the centre of the liquid. Streamline layers in the liquid will therefore exhibit continuously less flow motion from the centre of the liquid outwards. All three of these predictions are compatible with observation.

The atomic picture is readily extended to the turbulent flow, often encountered at high flow velocities down wide channels. The atomic kinetic energy is then used, not predominantly for local atomic motion, but to rotate the liquid as large eddies. The eddy motion is then very dependent on the density of the fluid (i.e. the mass of the rotating units). Furthermore, the fluid layers at rest on the vessel wall would not be expected to take part in the eddy motion. This is so in practice, where it is found that a boundary layer of streamlined flow exists at the container edges of the turbulent flow.

3.3.2. Non-Newtonian behaviour

Elastic materials which obey Hooke's law, are characterized by elastic moduli or constants. These moduli are the ratios of the appropriate stress applied to the corresponding strain produced. We have already seen that elastic moduli are a reflection of the interatomic restoring forces of Fig. 3.2(b). Similarly, the viscosity of a liquid represents its internal friction or opposition to local atomic displacement. It is not surprising therefore that η bears some resemblance to the elastic moduli of a solid. With η, because of the time dependence of fluid flow, one considers the *strain rate* ($\dot{\gamma}$) of the response. For viscous liquids under streamlined flow, with σ the applied shear stress, the coefficient of viscosity η is defined as

$$\eta = \sigma/\dot{\gamma}. \tag{3.1}$$

It is the liquid flow equivalent to the rigidity of a stressed solid. A liquid with constant η implies a linear response of $\dot{\gamma}$ to σ and this is reminiscent of Hooke's law. Materials which do not obey this linear relationship are called *non-Newtonian*. These materials are usually composed of large molecules of complicated shape. The deformation becomes more pronounced with increasing shear stress (Fig. 3.7). Typical of this class are polymers. At room temperature polymers can be crystalline (or amorphous) or liquid depending upon their molecular weight and composition. For low molecular weights (which are nevertheless still very high compared to other materials), the molecules in the polymer liquid become ordered rather than entangled as the shear stress is increased. Initially, this may involve the rupture of certain bonds, but once completed, the molecules elongate and glide over each other.

Gels and pastes are a common class of material in which the molecules are interlocked by relatively weak bonds. When subjected to relatively small shear stresses, these bonds are easily ruptured, and the gels collapse to a

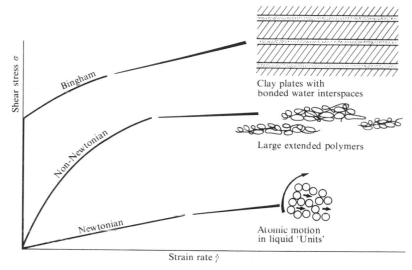

Clay plates with
bonded water interspaces

Large extended polymers

Atomic motion
in liquid 'Units'

Strain rate $\dot{\gamma}$

FIG. 3.7. Types of fluid flow, illustrated with representative materials.

liquid state which is regular in its flow behaviour. Such materials are called *thixotropic* and are found in gel paints. With time, the material regels upon removal of the stress.

Finally, *Bingham* flow is an interesting case. Here, relatively strong bonds maintain the stable structure. An example is clay in which discrete crystalline layers are interspersed with layers of water molecules. A high initial shear must be applied to separate the layers (Fig. 3.7). Once parted, these layers glide over each other, assisted by the flow of the released water. The requirement of a large initial stress enables a clay object to retain its shape against gravity before being fired in an oven.

3.3.3. *What about solids?*

Solids can also suffer flow deformation; it is simply a matter of the time scale involved. The motion of atoms into vacancies, used to explain the viscous deformation of liquids, is applicable to solids except that a rigid lattice is involved in the solid case. The interatomic forces impose great rigidity on the structure and hence allow the vacancies to be transferred within the interns of the material. However, under continued stress, the vacancies and dislocations break out of the solid surface and this marks the commencement of flow deformation. This effect is generally called *creep* in solids. Hence, one might consider that solids can be differentiated from liquids, not because of the absence of flow but rather through the magnitude of the viscosity coefficient.

3.4. Surfaces

3.4.1. *A single surface*

Within condensed matter, the atoms arrange themselves in positions of equilibrium which correspond to a balance of the interatomic forces consistent with the internal energy of the material. Let us suddenly cleave the material and separate the two fragments so formed. At the instant of cleavage the boundary atoms have a different number of close neighbours on one side and are no longer at equilibrium. These atoms move in order to establish a new equilibrium. In general this motion is towards the bulk.† The final boundary so formed is called a surface. Because the arrangement and relative positions of the surface atoms are now different from those in the bulk of the sample, we expect the surface to differ somewhat from the bulk in its properties. This picture assumes that atoms of the surrounding atmosphere have a negligible attraction for the surface atoms compared to that of the atoms in the bulk. This is generally true except for a liquid in contact with its vapour when conditions are close to the critical temperature of the substance.

In general, motion of the atoms towards the bulk in forming a surface is consistent with a reduction of the surface area of each of the so-formed fragments. With solids, the surface atoms cannot move far before meeting the rigid atomic lattice and entering the zone of the rapidly increasing repulsive forces (Fig. 3.2(b)) of the rigidly bonded inner neighbours. This will be true not only for a perfectly crystalline solid but also for amorphous materials, as the 'local' atoms near the surface will be highly ordered. In a liquid, there is considerably more freedom of motion, so that atoms and units can rearrange themselves and satisfy the tendency of the surface to reduce its area. The shape with the smallest surface area for a given volume is a sphere. Hence the liquid gathers into spherical drops.

If we wished to increase the surface area of a substance without the surface rupturing, atoms would have to be brought from the interior to constitute the greater surface. This would involve breaking bonds and hence performing work on the system. In other words, the surface of a material has greater energy than its bulk. The *excess* of energy over the bulk is called the *surface free energy*. It is applicable to both solid and liquid surfaces although its presence is more dramatically manifest in the case of liquids. For example, if we dip a capillary tube into a bath of liquid and hold it there, most liquids rise up the bore of the tube to a certain height above the liquid surface of the bath. This property, which initially appears to contradict the laws of hydrostatic pressure, may be explained in terms of the free energy of the open liquid surface in the capillary tube providing the energy required to maintain the raised liquid column. In fact, the complete explanation must take into account

---

† This behavior is not always encountered. In ionic crystals, such as those of the alkali halides, the atoms move outward to obtain equilibrium.

the free energy of the liquid–vessel surface (which gives rise to the angle of contact) as well as the liquid–vapour surface. This leads us to the next, more practical consideration.

3.4.2. Interfaces

Any surface must be in an environment. If this is anything other than a vacuum, the surface atoms experience forces due to atoms within its own bulk and those of the adjacent medium. On creating such a boundary between two media, the surface atoms move under the influence of these forces to form a state of equilibrium. The resultant boundary is termed an *interface*. If the interface results from preferential motion of surface atoms towards their own bulk medium, we speak of *cohesion*, whilst net movement towards the second medium is termed *adhesion*. Adhesion is simply the ability of one substance to bond to another. Let us briefly consider a few types of interface.

If we place a liquid on a solid surface, the liquid may either spread across the solid or gather into droplets. If spreading occurs, adhesion is the predominant factor, whilst drop formation indicates that cohesion is the important effect. A combination of these two effects is always present to some degree. For example, although water tends to gather into drops on a glass surface, these drops still cling to the glass.

Thus there is always some adhesion at a liquid–solid interface. This indicates that each atom in the liquid surface influences the arrangement of atoms in the solid surface and vice versa. Hence each substance mutually influences the surface free energy of the other material. For example, a liquid reduces the surface energy of a solid upon which it is placed. Solder action depends on this ability: the molten metal reduces the surface energy of a high-melting-point solid metal to which union is required. A liquid on the other hand can increase its surface area by spreading, thereby partially or completely restoring its own energy loss by getting closer to more solid atoms. The liquid continues to spread over the solid until the sum of the modified free energies of both liquid and solid surfaces is a minimum. We then speak of the 'free interfacial energy' of the system. In our reasoning, we have ignored the relatively weak interaction between either material and the gaseous or vapour atmosphere in which the system is situated. For complete understanding we ought to think of the energy of all three interfaces between solid, liquid, and vapour reducing to a minimum for equilibrium.

With two immiscible liquids the behaviour is similar to the above, except that both surfaces are able to readjust their shape. Gravitational forces also become important in that the more dense liquid will fall through the less dense if initially placed on the top. After that, both liquids will spread until the mutual interfacial energy becomes a minimum.

The solid–solid interface is of interest for two reasons. Firstly, it is possible to bond or 'sweat' certain solids together simply by pressure. The surfaces must be perfectly clean and smooth. The adhesion depends upon strong bonds being established between the surface atoms of each solid. Secondly, this mechanical world often depends on the ability of one solid to slide over another without interfacial linking. Whatever the incompatibility of the atoms in two touching surfaces, some weak bonding will occur between them, even if the surfaces could be perfectly smooth. This results in friction to motion. In the region of common contact, the free energy of each surface is changed. This explains to some extent why friction does depend experimentally upon the area of contact of the two sliding bodies, even though the elementary laws of friction state an independence of the common area.

Finally, the term *adsorption* is often used when describing the viscous flow of materials down tubes and channels. The atoms of a gas or liquid are attracted to a surface and become weakly bonded to it. They do not penetrate into the body of the host, and are called *surface active substances*. If they lower the free energy of the surface, 'positive' adsorption results. It is the tendency of gases and impurities to adsorb to metals and solids that prevents the easy preparation of very clean surfaces. Should the atoms of the fluid or impurity penetrate into the bulk of the host, the term *absorption* is used.

4. Bonds and bands

'If he went not through the narrow, how could he come into the broad?'

Apocrypha 2 Esdras 7; 5

4.1. From bonds to bands

In Chapter 2, we considered the types of bonds that could exist between individual pairs of atoms. In a condensed state, however, any atom finds itself in proximity with a number of close neighbours. This means that we must consider the cooperative effect of many bonds between many atoms if we are to account for the properties of actual materials. In practice, in a state of greatest order (i.e. zero entropy), the atomic binding within a system leads to that atomic array for which the internal energy of the whole structure is a minimum.

Electrons in atoms exist in discrete energy situations called *energy levels*. As atoms combine to form the condensed state, their individual atomic orbitals overlap and perturb the original energy levels of the isolated atoms. Now, the number of electrons that can occupy the resultant bonding orbital of the condensed state must depend on both the number (N) of interacting atoms and the type and number (M) of atomic energy levels involved. Basic atomic theory suggests that each atomic energy level houses a maximum of two electrons,[†] thus the final bonding situation is clearly capable of accommodating $2 \times N \times M$ electrons. The bonding orbital for the condensed material consists of a series or *band* of $N \times M$ closely spaced energy levels (Fig. 4.1). Since these bands originate from the energy levels of the individual atoms, we speak of a *splitting* of atomic levels into energy bands.

Two points are worthy of notice. Firstly, the newly formed bonding orbital represents a minimum-energy condition, and hence the band of energy levels will correspond to a lower electronic energy situation than that of the original atomic levels (Fig. 4.1). Secondly, as the splitting of the atomic levels originates in the interaction between atoms, the degree of splitting is a measure of the atomic interaction. The outer valency levels interact much more strongly than the inner-core levels. Hence the valency levels split into relatively wide bands whilst the core splitting is very narrow. These narrow bands are little reduced in energy from the original atomic levels (Fig. 4.1).

In the literature, bond theory may be used to describe the properties of one material class, and band theory another. The reader should not adopt the view that these are contradictory approaches. We have shown above that bond

[†] The limit of two electrons is a consequence of the statistics discussed in Chapter 6.

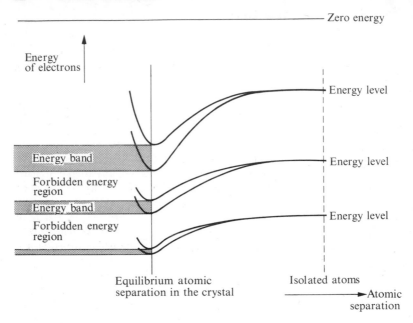

FIG. 4.1. The formation of energy bands by the splitting of atomic energy levels. In this and subsequent diagrams energy values are negative, measured relative to an energy zero at the top of the diagram.

and band theories are compatible. The band theory does have special versatility however, when applied to the electrical properties of solids, and in this context generally it is used exclusively.

4.2. Band characteristics of materials

Historically, real crystalline materials were divided into two classes according to certain of their physical properties. Crystals which were optically highly reflecting and also good electrical conductors were called *metals*. They included group I and group IIA elements of the periodic table. Others, which were transparent at certain wavelengths in or near the visible region of the electromagnetic spectrum and were bad electrical conductors were termed *insulators*. They consisted of elements in the top right hand corner of the periodic table, particularly those from groups IVB and VIB, together with a number of ionic and covalent compounds. The original development and success of band theory was to supply a means of accounting for the different properties of these classes at an atomic level. In this section we shall confine the discussion to the conduction properties of ideal crystals. Less ordered structures will be considered later.

Electronic conduction results from a net motion of electrons under the influence of an applied electric field. The existence of a finite potential difference (p.d.) suggests that energy is required to maintain a current flowing in a material. This energy measures the degree to which the material resists a change in motion of its constituent electrons. Hence the ratio of the applied p.d. to the resultant current is termed the *resistance* of the material. To characterize materials it is better to use the *electrical resistivity* (ρ) defined as the resistance of a 1 m cube of the material (measured in ohm m). The reciprocal of this quantity is called the *electrical conductivity* (σ). Thus materials with high conductivities should be those in which large numbers of electrons are available, whose velocity (v) and momentum ($m_0 v$—where m_0 is the electron mass) may be easily changed. As we shall see in Chapter 5, changes in energy and momentum are related, and the conductivity will depend on the electronic energy spectrum and the distribution of electrons within this spectrum. The remainder of this chapter is concerned with the influence of the energy spectrum on conduction.

4.2.1. *Good conductors*

Consider the band model of an ideal crystal. We recall that bands are derived from the splitting of atomic levels. In a crystal containing N atoms, each atomic level leads to a band capable of containing $2N$ electrons. Now, in an atom the deep-lying electronic levels generally contain two electrons so that their resulting bands contain $2N$ electrons and are filled. As the energy bands are derived from the atomic levels of the member atoms, they maintain to some degree the *forbidden energy regions* of the atoms (Fig. 4.1). As we shall see later, this is strictly only true for an ideal lattice. Thus the inner-core electrons have no readily available energy levels within their band into which they may be excited and hence they cannot contribute to electronic conduction. Since their band is full, any change in energy and velocity for one electron is compensated by equal and opposite changes in these properties for another electron. Thus we cannot observe their motion. It would appear that the inner electrons do not explain the difference between conductors and insulators, so we must look carefully at the valence electrons for this.

Many typical metals lie in groups I and II. Single atoms of the group I metals possess valence shells containing one electron. This electron is normally situated in an atomic level capable of containing $2N$ electrons. The band is thus only half-filled (Fig. 4.2). Hence there are levels within the band into which electrons may be excited. By supplying external energy we may excite electrons into higher vacant energy levels within the band (e.g. a transition from A to B in Fig. 4.4(a)). The energy change is not compensated by another electron losing energy. Hence, charge may be observed to move and a current may be induced to flow through the material. The system thus has the ingredients of a good conductor.

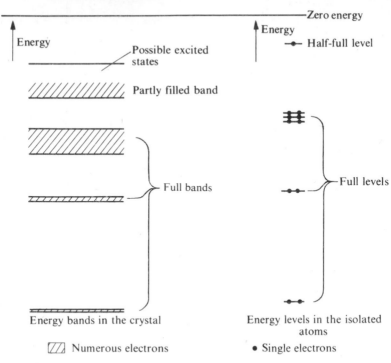

FIG. 4.2. Schematic energy spectra for electrons in crystalline and atomic sodium—a typical group I metal.

Group II elements are also good conductors. The valence shell consists of two electrons in a filled atomic level. The alert reader may point out that the filled levels of these elements should lead to a completely filled outer band, with no likelihood of conduction. The answer to this apparent anomaly lies in the concept of hybridization introduced in section 2.1. With this phenomenon, more than one atomic level per atom may contribute to a bonded state. In fact, for the group II metals, empty atomic levels in the valence shell give rise to a corresponding empty energy band in the crystal. The adopted structure of group II metals is such that certain empty and full bands may overlap, forming one single 'hybridized' band (Fig. 4.3). Clearly, this band is only partly filled with electrons and electron conduction is possible upon provision of external energy to excite electrons.

4.2.2. *Insulators*

What then makes a material an insulator? The answer lies in the fact that a filled energy band has little potential for conduction. A number of

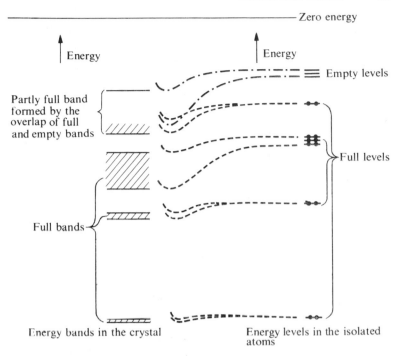

——————————————————— Zero energy

↑ Energy   ↑ Energy

Partly full band formed by the overlap of full and empty bands

Full bands

Energy bands in the crystal

Empty levels

Full levels

Energy levels in the isolated atoms

▨ Numerous electrons   ➤ Single electrons

FIG. 4.3. Schematic energy spectra for electrons in crystalline and atomic magnesium —a typical group II metal.

non-metallic elements such as those from groups IVB and VIB, bond by donating or sharing electrons, so as to achieve filled valence shells. In the crystalline state, the outermost band, called the *valence band* since it is composed of levels formed from the valence electrons, will thus be completely filled in these materials (Fig. 4.4(b)). Conduction would thus appear to be impossible and we have an insulator.

One problem still remains. Although they are not good conductors, insulators can be made to conduct. Why do they conduct at all? Above and removed from the filled valence band, there is another band of allowed energy levels called the *conduction band* (Fig. 4.4(b)). These two bands are separated by an energy gap. For electrons to conduct, we must supply energy to free them from their bonds or, on the band picture, to excite them into the conduction band (e.g. a transition from C to D in Fig. 4.4(b)). They then have accessible energy levels and may conduct in an electric field. Only in very rare cases can an electric field itself provide sufficient energy to excite electrons.

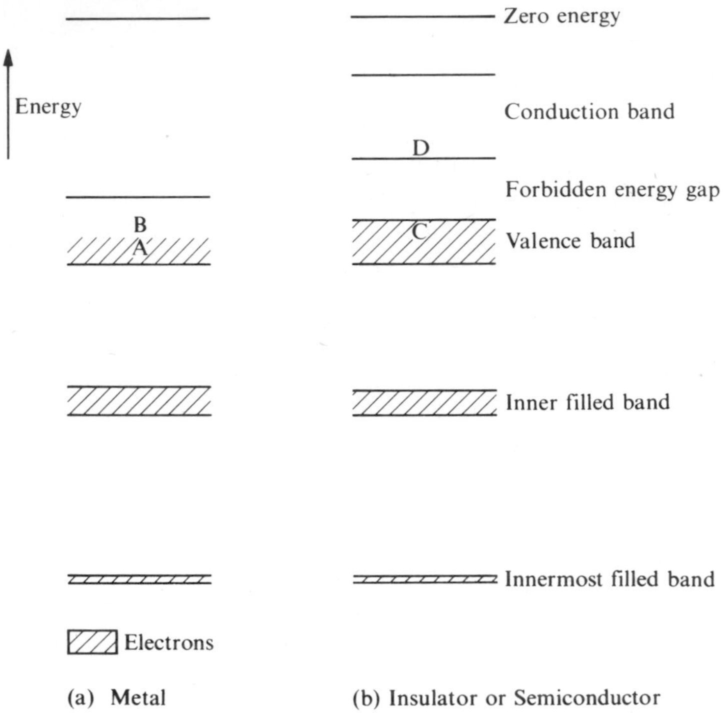

———————————— Zero energy

Energy

———————————— Conduction band

D ————

Forbidden energy gap

B

A

C

Valence band

Inner filled band

Innermost filled band

Electrons

(a) Metal   (b) Insulator or Semiconductor

FIG. 4.4. Band pictures for metals and insulators or semiconductors.

In most cases another external agency must be introduced. The commonest source of excitation in a real crystal is thermal excitation. However, the number of electrons freed, and hence the magnitude of the electronic conductivity, clearly depends on the relative magnitudes of the energy available for excitation (E_T) and the binding energy (E_b) of the electron. If E_b is large with respect to E_T then only a few carriers are produced and the material is considered as an insulator. If, however, E_b is sufficiently small, such that E_T may free a large number of electrons, then the material behaves like an insulator but with increased electrical conductivity. Such materials, which then strive to ape true conductors (metals), are called *semiconductors*.

In Chapter 2 we mentioned that materials may be anisotropic, showing different strengths of bonding with various positions and directions within the crystal. In this case the binding energy or band gap (between valence and conduction bands) will also vary with direction. Similarly the electronic conductivity will vary with direction in the crystal. One·extreme possibility is that the band gap, although non-zero in some directions, may become zero in others. A good example is graphite, where the ratio of its directional conductivities may be 10:1 for suitably chosen directions in the crystal. Materials with very small amounts of band overlap are called *semi-metals*.

4.3. Periodicity

Previously in this chapter we have discussed mainly the ideal crystal. We have permitted external factors to influence only the excitation of electrons from their normal states. Now, an important property of an ideal lattice is that it can be constructed by repeating a particular atomic arrangement, the unit cell of section 2.2.1, in three dimensions. The lattice is therefore periodic. Those electrons in a conductor which reside in the conduction band and move under the influence of an electric field, are called *conduction electrons*. In an ideal crystal they are completely 'free' and roam anywhere throughout the crystal. As we shall see in the next chapter, an elegant mathematical model has been developed which embraces the electronic energy picture of the crystalline lattice and includes the concept of the forbidden energy regions. We note here that the model pictures the freedom of the conduction electrons as a direct consequence of the crystal periodicity. Loss of perfect periodicity is thus expected to lead to noticeable changes in the conduction properties of a material. In fact, loss of order and hence periodicity may affect the conductivity in two ways.

Firstly, the multiplicity of defects listed in Chapter 2 all lead to a loss of periodicity and restrict the motion of electrons through the material. They thus contribute to the resistance of the material. This type of effect will be discussed further in Chapter 5.

Secondly, certain specific defects may give rise to a change in the electronic energy spectrum and hence affect the supply of conduction electrons. This property is easier to explain by way of an example. Imagine a covalently bonded lattice comprised of group IVB atoms. Through bonding, each atom gains an inert structure. Let us consider the effect of replacing any host atom by a group VB element. After bonding, this element will still possess an extra electron. This electron is not required for bonding. It is much easier to free from its parent site than any of the 'bonding' electrons. Hence the group VB atom is effectively an impurity in the lattice, which provides an additional energy level, spatially localized within a few atomic radii around the impurity. Such levels occur within the forbidden energy region of the impurity-free crystal (Fig. 4.5). They provide easily available electrons and are called *donor levels*. If instead of a group VB element we introduce a group IIIB element, then on the band picture there is a deficiency of one electron localized at our impurity. Such a structure invites electrons to leave their bonding states and fill the deficiency. We term these *acceptor levels*.

In general, any defect will disrupt the crystal periodicity and introduce localized levels into the forbidden energy regions. We might emphasize the

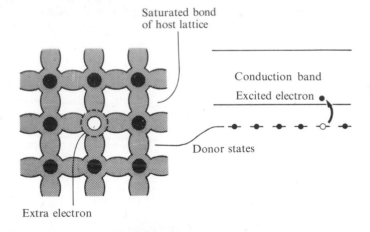

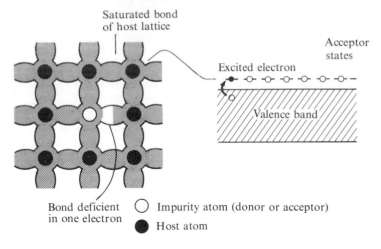

FIG. 4.5. Bond and Band pictures of donor and acceptor states.

importance of impurities by considering the effect of donor impurities in silicon. Highly purified silicon may be made with intrinsic conductivities as low as 5×10^{-4} (ohm m)$^{-1}$ at room temperature. The introduction of say 10^{24} donor/m^3 leads to conductivities as high as 10^2 to 10^3 (ohm m)$^{-1}$ at room temperatures. For a typical atomic density of 10^{28} atoms/m^3 we see that this 10^6- to 10^7-fold increase in conductivity results from the influence of only one part per ten thousand donor atoms.

4.4. Amorphous solids

We began this chapter by considering the properties of an ideal crystal, and then considering the effects of introducing a small number of defects. Suppose we allow the degree of disorder to increase to such an extent that we no longer have a periodic structure with defects, but an essentially non-periodic structure which has local ordered regions. We recognize such structures as the glassy or amorphous materials described in section 2.2.2. Experimentally there is evidence for electronic conduction in such systems. Furthermore, there seems to be some practical justification for classifying these materials into the two classes of 'metals' and 'semiconductors'.

In order to explain this difference between the metallic and semiconducting behaviour in amorphous materials, it is tempting to try and retain the band model. To this end it is useful to consider the effect of increasing the density of defects present in a crystalline lattice.

In general the band structures of metals are much the same in form in the crystalline and amorphous states. One exception arises with metals such as those in group IIA, in which the metallic behavior depends on the overlap of filled and empty bands (Fig. 4.3). On melting, the average atomic separation changes, the overlap of the bands decreases, and the melt shows a tendency to become more like a semiconductor with a small energy gap.

The effect on crystal semiconductors of introducing defects of various types is, however, more noticeable. For this reason we shall consider progressively disordered semiconductors in some detail. We have already seen the effect of small densities of certain impurities in a crystal semiconductor

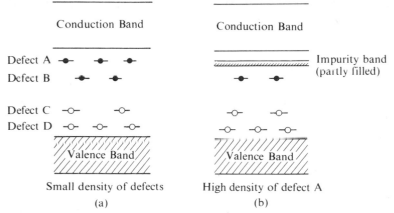

FIG. 4.6. Impurity levels in semiconductors. (a) Isolated defects at low doping levels. (b) High density of defect A broadens the once isolated donor levels into an impurity band.

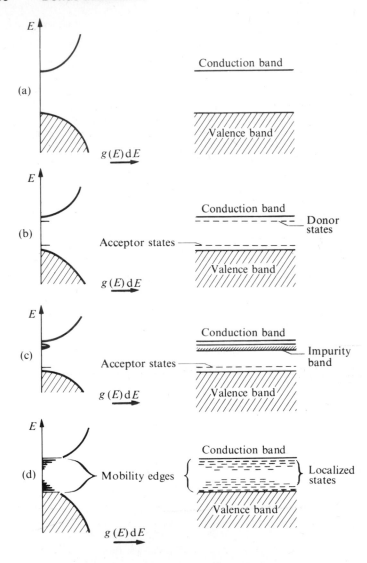

F IG. 4.7. The effect of disorder on a crystalline semiconductor illustrated in terms of the conventional band diagrams and plots of $g(E)\,dE$ the number of levels contained in the energy interval E to $E+dE$. (a) Completely pure crystal. (b) Lightly doped crystal. (c) Crystal heavily doped with one type of impurity. (d) Amorphous semiconductor showing the mobility edges and the high densities of localized levels. (Note: energy values are negative, increasing to zero at the top of a figure.)

(section 4.3). One picture is to allow the density of one type of defect to increase. When the effective density of one such defect (defect A in Fig. 4.6) exceeds approximately one part in ten thousand, these defects become sufficiently close spatially to cause a splitting of the individual localized levels into a band of levels called an *impurity band* (Fig. 4.6(b)). Thus conduction may occur by the motion of electrons between the defect sites without the need to activate such electrons either into the conduction band or from the valence band of the host material.

However, we are still dealing with relatively small disruptions of the crystalline lattice, and thus the periodic nature of the host lattice is still maintained. In a truly amorphous material the structure has been sufficiently perturbed to destroy the long-range order in the system. However, provided the short-range order resembles that of the equivalent crystal, over distances greater than the unit cell dimensions, then the nature of the local bonding is not greatly changed. In section 4.1 we pictured the band model as arising from the binding of the atoms in forming crystals. For any particular atom within the crystal, its nearest-neighbour interactions are the most important. Hence we might expect the essential properties of the band approach to be applicable to amorphous systems.

The disordered nature of the amorphous state will obviously affect the detail of the band structure. In fact the proliferation of defects is believed to lead to a large number of localized levels, of various energies, in the once forbidden energy regime (Fig. 4.7(d)). For certain amorphous semiconductors the localized levels may completely fill the energy region between the bands. Electron motion amongst localized sites may occur but this is more difficult than the motion of electrons in either the conduction or valence bands. Electrons moving within the bands are more mobile and thus the energy boundary between the bands and the localized states are referred to as *mobility†ldots edges* (Fig. 4.7(d)). Thus the 'bands' are now separated by a mobility gap rather than an energy gap.

Thus in this chapter we have viewed the band structure of the perfect crystal and considered the effects of introducing an increasing amount of disorder. In the case of a semiconductor the effect of increasing disorder is summarized in Fig. 4.7.

---

† The mobility (μ) of electrons in a material may be defined as the ratio of the average velocity (v) imposed on the electrons by an applied electric field, to the magnitude (E) of this electric field: that is $\mu = v/E$.

5. Electronic motion in matter

'Electricity is of two kinds, positive and negative. The difference is, I presume, that one comes a little more expensive, but is more durable; the other is a cheaper thing, but the moths get into it.'

STEPHEN LEACOCK *Literary lapses—A manual of education*

I N the previous chapter we have indicated that analysis of electronic motion within a solid is concerned with how electrons may change momentum and energy. This involves discussion of the normal and excited electronic energy levels of the solid and the distribution of electrons within them. The band model was introduced as it provided a clear distinction between metals and insulators (or semiconductors) and could be readily related to the chemical bonding. However, the band model is not the only picture of the electronic behaviour. For completeness, we ought briefly to mention the older historical models. Originally, as metals were studied more avidly than insulators, early models were aimed primarily at accounting for the properties of crystalline metals. In the first section of this chapter we shall follow the development of these models and then later discuss their relevance to current knowledge of the properties of both crystalline and amorphous materials.

5.1. Metals in the crystalline state

Metals are known for their high electrical and thermal conductivities. Such properties have always been associated with the apparent freedom of electrons to move through the crystal lattice. Models of the metallic state have centred around this freedom of motion.

5.1.1. *Classical picture*

This was enunciated soon after the discovery of the electron. At that time the kinetic theory of gases was well established. In this model the molecules of a gas are considered to have independent, random motion within a container and to be subject to many collisions with each other and the vessel. It was a natural extension to view the valence electrons as a cloud of negatively charged particles, which frequently collide with the vibrating lattice. Such collisions lead to an exchange of momentum and energy between the lattice and the electron cloud. The net momentum and energy gained by the electrons is thus a balance between the change in energy and momentum due to external energy sources and the changes in these quantities due to lattice collisions. The picture of the electrons as an analogous 'gas' suggests that all the valence electrons might be capable of gaining energy and momentum; these quantities being distributed somewhat equally amongst these electrons by means of their collisions with each other and the lattice. Thus introduction of energy from an external source (e.g. raising the temperature) raises the average energy and hence the kinetic energy of the majority of the electrons. This is illustrated in Fig. 5.1(a) where the movement of the peak of the curve shows the change in average kinetic energy as the temperature is raised. Such a distribution of energy amongst interacting particles is termed a *Maxwell–Boltzmann distribution* and is characteristic of the classical approach. The reader may recognize each curve of Fig. 5.1(a) as that of the distribution of molecular kinetic energies in a gas according to the kinetic theory. The curve has two noteworthy features.

Firstly, at a given temperature the kinetic energy of an electron, E_k, is given by

$$E_k = \tfrac{1}{2}m_0 v^2 \tag{5.1}$$

for an electron of mass m_0 and velocity v. We may rewrite this equation in terms of the electron momentum ($p = m_0 v$). Hence

$$E_k = \tfrac{1}{2} \cdot \frac{p^2}{m_0}. \tag{5.2}$$

Hence, a graph of E_k versus p is a parabola (Fig. 5.1(b)). Introduction of energy increases E_k and allows the electron to move up the parabola. The final position of the electron on the graph depends on the balance between the energy supplied and that lost by collisions with the lattice.

Secondly, the average kinetic energy is found to be proportional to the kelvin temperature. By analogy with the kinetic theory of gases, the heat stored by the electrons is often indicated by an 'electronic specific heat'. If an amount δQ of heat is supplied to a unit mass of the metal and leads to an increase δT in the temperature, then the electronic specific heat C_e is defined as

$$C_e = \underset{\delta T \to 0}{\mathrm{Lt}} \left(\frac{\delta Q}{\delta T} \right) = \frac{\mathrm{d}Q}{\mathrm{d}T}. \tag{5.3}$$

The relatively large change in position with temperature of the peak of Fig. 5.1(a) suggests that C_e is large. The total electronic specific heat depends on the number of electrons in the cloud. Now an insulator has a relatively low density, and a metal a high density of electrons which are available for conduction. Hence, the metal should possess a much higher specific heat than the insulator, assuming that the lattice specific heats are similar. In practice, metals and insulators have similar total specific heats. This fact, together with the inadequacy of the theory to differentiate between conductors and insulators (plus certain other difficulties associated with the magnetic properties of the materials) led to the ultimate abandonment of the classical model.

5.1.2. *Free-electron theory (F.E.T.)*

In the classical approach the electrons were pictured as a cloud or sea of particles throughout the body of the material. The only property of the lattice

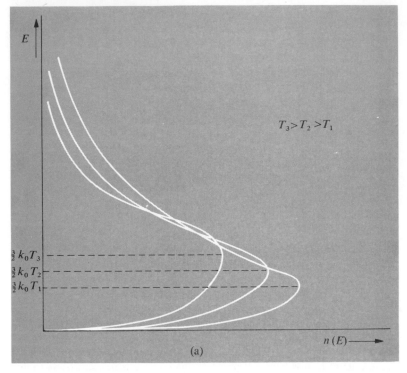

E

$T_3 > T_2 > T_1$

$\frac{3}{2} k_0 T_3$

$\frac{3}{2} k_0 T_2$

$\frac{3}{2} k_0 T_1$

$n(E) \longrightarrow$

(a)

FIG. 5.1. Classical model of the energy distribution amongst electrons in a metal. (a) Maxwell–Boltzmann distribution. The average kinetic energy at temperature T corresponds to the peak of the curve.

was that it provided collision points for the random motion of the electrons. The failings of the classical approach led to a new picture in which the electrons in a metal were still visualized as being free from any particular lattice 'atom', but were collectively bound to the lattice. The fact that energy is needed to remove electrons from a metal was attributed to the general and *communal* interaction of all the electrons with the positively charged ionic lattice. The interaction was pictured as uniform throughout the crystal.

The energy spectrum for such 'free electrons' may be calculated by modern quantum mechanical methods. The resulting energy levels are shown to be a series of closely spaced levels forming an energy band. However, by comparison with the classical picture, the distribution of electrons amongst these levels is startingly different (Fig. 5.2(a)). Even at absolute zero temperature the electrons would completely occupy levels up to a finite value E_F (the *fermi energy* or *level*). Astonishingly, for metals, if E_F is considered as the average kinetic energy of a 'classical gas of electrons' then the temperature of the electron

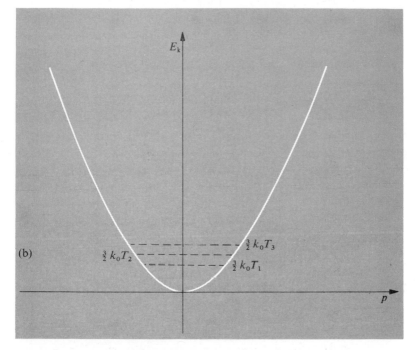

FIG. 5.1. (b) The parabolic relation between the electron kinetic energy and the momentum p. At temperature T, the average kinetic energy is $3/2k_0 T$ where k_0 is called the Boltzmann constant.

cloud would be approximately 50 000 kelvins. Clearly, most external sources of energy will supply quantities of energy much less than E_F and hence lead to only small changes in the electron energy distribution (Fig. 5.2(a)). The complete distribution curve is determined by the fact that the electrons obey a form of statistics known as Fermi–Dirac statistics. (See Chapter 6.) This is schematically represented in Fig. 5.2(b). The curve indicates the probability $f(E)$ of an energy level being occupied. Thus the number $n(E)\,dE$ of electrons actually present in the energy levels contained in the range of energy between E and $E + dE$ (Fig. 5.2(d)) is governed by the product of three factors. These are: the probability of occupation $f(E)$, the number $g(E)$ of available levels per unit energy interval and the energy interval dE (Fig. 5.2(c)).

The electrons are still free and hence the curve of electron energy versus momentum is still parabolic. However, at absolute zero of temperature the energy of the electron need not be zero but may have any value up to E_F. Further, if energy is supplied from an external source, then electrons can accept it only if excited levels are available into which they may be activated.

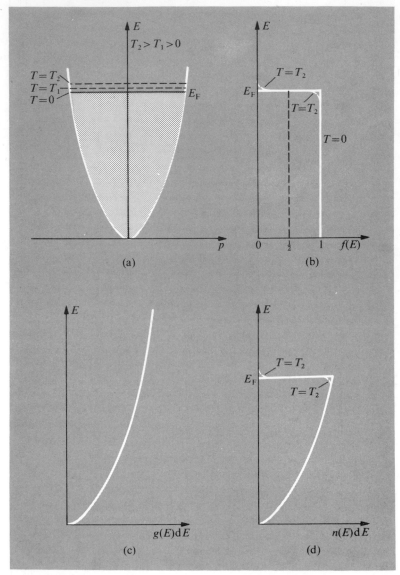

FIG. 5.2. Free-Electron Theory of the distribution of energy amongst electrons in a metal. (a) The parabolic relationship between electron energy and momentum. (b) The Fermi–Dirac statistical distribution, indicating the probability of an energy level being occupied. Values of $f(E) = 1$ guarantee full occupation whereas $f(E) = 0$ forbids occupation. (c) The distribution $g(E)\,dE$ of available electron energy levels. (d) The number of electrons $n(E)\,dE$ actually occupying available energy levels. Shaded area represents levels occupied by electrons.

Let the quantity of external energy supplied be δE. If, as is often so, δE is very small with respect to E_F, then only those electrons having energies greater than $(E_F - \delta E)$ stand any chance of being excited. Hence storage of external energy by electrons is restricted to a relatively small fraction of the total electron population. Thus the total electronic specific heat of the metal is small, and one difficulty of the classical method is removed. However, there still remains the problem of distinguishing between metals and insulators. In fact, the free-electron model would suggest that *all* elements must possess partly filled bands and hence exhibit metallic behaviour.

In conclusion, the free-electron model accounts reasonably well for metals but is unable to explain the properties of insulators.

5.1.3. *The band theory*

In the free-electron theory, the influence of the lattice was considered to be uniform throughout the material. The extension of the F.E.T. which led to the band theory was to consider the effect of position within the lattice on the electron–lattice interaction. Crudely put, the electrons are attracted by the lattice to a different degree in the vicinity of lattice 'ions' or 'atoms' than in the interspaces.

Now an important property of a crystalline structure is the periodic array of the lattice sites mentioned in the previous chapter. Hence, the electron–lattice interaction is considered as periodic, rather than constant as in the F.E.T. A quantum mechanical analysis of this situation yields an electron energy spectrum which is similar to that of the F.E.T. but with the addition of forbidden energy regions (Fig. 5.3(a)). Such allowed and forbidden regions are the bands and gaps introduced in Chapter 4. Figs. 4.1 and 4.2 show plots of energy against position within a crystal whereas Fig. 5.3(a) plots energy against momentum for a given position in the crystal. The electron motion (momentum) is restricted to those values which correspond to energies within these energy bands. No transport may occur for electron momenta (and hence energies) corresponding to the forbidden energy conditions. Clearly, the complete parabolic relationship of eqn (5.2) is not strictly obeyed, although the outline of the original parabola can still be identified in Fig. 5.3. In many cases it is possible to account for the behaviour of electrons within bands by means of an equation of the type (5.2) provided one replaces m_0 by m^*, an effective mass which takes account of the deviation of the true E–p relation from that of an ideal parabola.

Thus we see that the periodicity of the ideal lattice, which we visualized in Chapter 4 as a consequence of the chemical binding and hence the forces between the lattice atoms, provides an origin for energy bands in real crystals. This permits a distinction to be made between insulating and metallic behaviour. To date the band theory is the best model for describing the behaviour of crystalline materials.

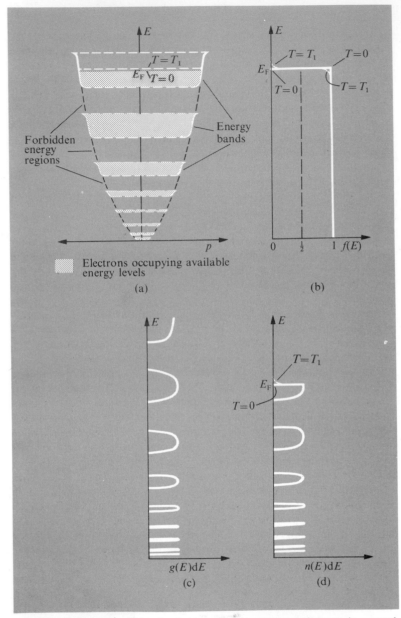

FIG. 5.3. Band model of the distribution of energy amongst electrons in a metal. The diagrams show: (a) The 'nearly parabolic' relation between energy and momentum and the existence of bands and forbidden energy regions. (b) The Fermi–Dirac distribution $f(E)$. (c) The number of energy levels $g(E)\,\mathrm{d}E$ available at various energies. (d) The number of electrons $n(E)\,\mathrm{d}E$ actually occupying the available energy levels.

5.1.4. *Metallic conductivity*

Two important properties of metals are their high electrical and thermal conductivities.† Now the former involves the transport of charge and the latter the transport of heat through a lattice. Let us suppose that electrons are responsible for both and note that finite potential or thermal gradients are necessary to maintain flow.

Now at a finite temperature T (kelvin) we introduce energy (approximately k_0T where k_0 is the Boltzmann constant) to the electrons and change their energy distribution in the range

$$(E_F + k_0 T) > E_F > (E_F - k_0 T). \tag{5.4}$$

We shall see that the electrons within this range play an important role in conduction phenomena. It is instructive to note that, in this range of energies, the F.E.T. and band model are similar (compare Figs 5.2 and 5.3).

In the absence of a potential difference (p.d.) the net momentum of all electrons is zero: there are as many electrons moving in any one direction as in any other. Applying a p.d. establishes an electric field which attempts to accelerate all the electrons in one direction. All electrons experience changes in momentum and energy due to the applied field. For those electrons deep down in the band where $E < (E_F - k_0 T)$, any gain in energy by one electron is balanced by the loss of energy by another electron. (No holes appear in the energy distribution.) However, electrons in the range given by eqn (5.4) may be scattered by the lattice and lose energy. Thus energy must be continually supplied to maintain a constant motion of charge. This is the phenomenon of electrical resistance and explains the need for a finite p.d. Clearly the electrical conductivity σ depends on the total charge (ne—where n is the number of electrons each of charge e in the band) and the mobility (μ) of electrons with energies near E_F. This mobility depends on the type of scattering mechanism and the degree of attachment of the electrons to the lattice measured by means of m^*.

In practice‡

$$\sigma = ne\mu. \tag{5.5}$$

The thermal conductivity (κ) depends on the transport of heat through the material. We shall be concerned with the electronic contribution to this phenomenon. In the absence of a temperature gradient, the net energy and momentum of all the electrons is zero. Applying a thermal gradient causes a distribution of electronic energies in excess of E_F along the temperature gradient. Electrons at the hot end of the sample possess greater energy and

---

† Thermal conductivity may be defined, by analogy with electrical conductivity, as the reciprocal of the thermal resistance (the resistance to the flow of heat) of a 1 m cube of material.

‡ Rigorous derivations of equations (5.5) and (5.6) can be found in the texts on solid-state physics in the bibliography.

are less easily scattered. Their motion allows them to pass energy by means of collisions to the electrons in colder regions of the material. Thus energy is conducted through the sample along the temperature gradient. Since some of the energy supplied by heating the material will be lost by the electrons to the lattice, a finite thermal gradient is necessary to maintain the flow of heat.

The thermal conductivity thus depends on the number of electrons involved in conduction, their mobility and their average energy in excess of E_F. In fact

$$\kappa = A_\kappa(k_0 T)n\mu \tag{5.6}$$

where A_κ is approximately constant for all metals.

It is interesting to consider the ratio

$$\frac{\kappa}{\sigma T} \simeq A_\kappa \frac{k_0}{e} \tag{5.7}$$

which is approximately constant and is independent of temperature. This relationship between the thermal and electrical conductivities is known as the *Wiedemann–Franz–Lorenz* law, obedience of which signifies the dominance of electronic contributions to electrical and thermal transport. Most metals and alloys obey this law over a wide range of conditions. Non-metals do not obey this law.

5.2. Crystalline insulators and semiconductors

Crystalline insulators and semiconductors are noted for their comparatively low electrical conductivities, their characteristic temperature *dependence* of conductivity, their optical transparency at visible or near-visible wavelengths and a surprising sensitivity of their properties to the presence of impurities or defects.

The band model was developed mainly to account for insulators and semiconductors, so that it is not surprising that it is predominantly used to describe the behaviour of these materials. To avoid the constant reference to both 'insulator' and 'semiconductor', we shall emphasize the difference between them and henceforth speak only of semiconductors. The difference is due solely to the size of the band gap $(E_c - E_v)$ between the conduction and valence band edges, as shown in Fig. 5.4. If $(E_c - E_v)$ is large with respect to the externally supplied energy, then the material is an insulator. For a crystal at room temperature, the vibrational energy available to release electrons from bonded states is of the order of $k_0 T$, that is about 4×10^{-21} joules (or $1/40$ electronvolts). Insulators have band gaps at room temperature some two hundred times this figure, that is in excess of 8×10^{-19} joules. Semiconductors have non-zero band gaps less than this figure at room temperature.

Thus, theoretically, all insulators can be semiconductors at high enough temperatures when the vibrational energy is large. In practice, the material

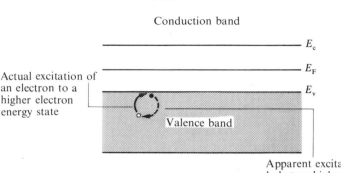

FIG. 5.4. An intrinsic semiconductor. Apparent motion of a hole in a nearly filled valence band. Note that motion to a high 'hole' energy level corresponds to motion to a low 'electron' energy level. E_c (the conduction-band edge) is the lowest energy level in the conduction band and E_v (the valence-band edge) is the highest energy level in the valence band.

may inconveniently melt or otherwise change phase before a suitable temperature has been reached.

5.2.1. *Intrinsic conduction*

With semiconductors, the availability of electrons for conduction depends on the freeing of electrons from bonded states and hence their excitation from the valence band into the conduction band. However, when we remove electrons from the valence band this band is no longer completely filled, and excited levels become available for other electrons within this band. We picture such absent electrons as leading to *holes* or free levels in the valence band. When an electron of lower energy in the valence band is excited so as to fill a hole, a new hole appears at the original energy level 'vacated' by the newly excited electron. It is often more convenient to picture the motion of the hole rather than that of the electron. The two are related schematically in Fig. 5.4. The motion of holes is equivalent to the motion of electrons provided one remembers that the holes act as if they have a positive charge. Free electrons in the conduction band may be excited to higher electronic energy levels. However, free holes in the valence band are *excited* to higher hole levels in the valence band which correspond to lower electronic energy levels. Thus, freeing an electron from a bond appears to generate two current carriers, a positive hole and a negative electron. Both are mobile and contribute to conduction. Since these current carriers are both provided by atoms of the host material, we speak of *intrinsic conduction*.

In a metal we pictured the fermi level E_F as the highest occupied energy level at absolute zero of temperature. Alternatively one can define the fermi level at zero or non-zero temperatures as that energy level within the material at which $f(E) = \frac{1}{2}$. These definitions are equivalent for a metal (see Figs. 5.2 and 5.3) but not for a semiconductor. For a semiconductor, one must define the fermi level as the energy for which $f(E) = \frac{1}{2}$ at non-zero temperatures. Then, the *total* number of electrons in the conduction band is dependent on both the number of *available* energy levels (Fig. 5.5(b)) and their probability of being occupied (Fig. 5.5(c)). It is thus equal to the area A in Fig. 5.5(e). Similarly the *total* number of holes in the valence band depends on the number of electronic levels which *may* be emptied (Fig. 5.5(b)) and the *probability* of actually emptying these levels (Fig. 5.5(d)). Hence area B in Fig. 5.5(f) represents this total hole density. Now, remembering what this density of electrons (and holes) signifies we see that area A represents the total free negative charge and area B the total free positive charge present in the lattice. In an intrinsic semiconductor holes and electrons are always produced in pairs. Hence we expect charge neutrality to be maintained in the lattice. Thus the areas A and B are equal. Further, let us stipulate that the holes and conduction electrons have equal effective masses ; that is, we picture both types of carriers as equally mobile. This means that the density of electronic energy levels at the bottom of the conduction band is the inverse of the density of the electronic energy levels at the top of the valence band. Thus area A is not only equal to area B but is also identical in shape. From the symmetry of the Fermi–Dirac distribution (Fig. 5.5(c), or (d)) we thus visualize the fermi level E_F as existing midway between E_c and E_v (Fig. 5.5(a)).

Now, suppose that the effective masses of the electrons and holes are different. This implies that their mobilities are different as a greater mass indicates lower mobility. Moreover we have already stated in section 5.1.3 that m^* reflects the shape of the energy versus momentum curve. Hence, the shapes of the densities, $g(E) \, dE$, of the electronic energy levels at the edges of the conduction and valence bands will differ as shown in Fig. 5.6(a). Because of this, the *shapes* of the areas A and B are no longer identical, even though their *areas* are still equal. This means that the fermi level, E_F, is displaced from the mid-point energy $\frac{1}{2}(E_c + E_v)$. Thus the position of the fermi level is sensitive to the effective masses and hence to the mobility of the charge carriers. In practice, it seldom corresponds to the mid-point value. Its actual position is useful in predicting the suitability of materials in the construction of electronic devices. We shall meet this in section 5.4.

5.2.1. *Extrinsic conduction*

In Chapter 4 we mentioned that donor and acceptor levels may be introduced into crystalline semiconductors. We recall that donor levels are close to the edge of the conduction band with the result that electrons may be

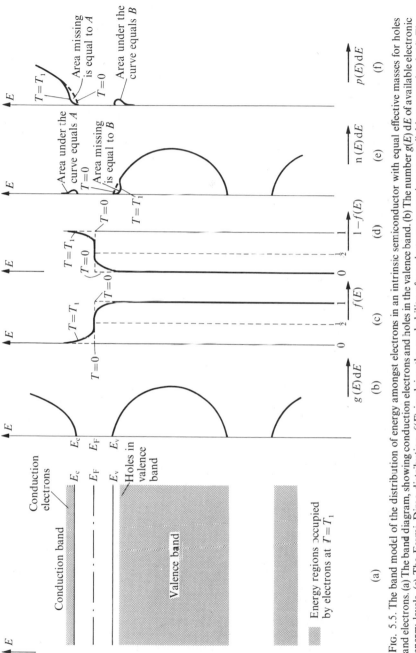

FIG. 5.5. The band model of the distribution of energy amongst electrons in an intrinsic semiconductor with equal effective masses for holes and electrons. (a) The band diagram, showing conduction electrons and holes in the valence band. (b) The number $g(E)\,dE$ of available electronic energy levels. (c) The Fermi–Dirac distribution $f(E)$ implying the probability of an electron occupying an available energy level. (d) The probability $1 - f(E)$ of an electronic energy level being empty. Hence diagrams (e) and (f) represent respectively the number of electrons present $n(E)\,dE$ and absent $p(E)\,dE$ from the available energy levels.

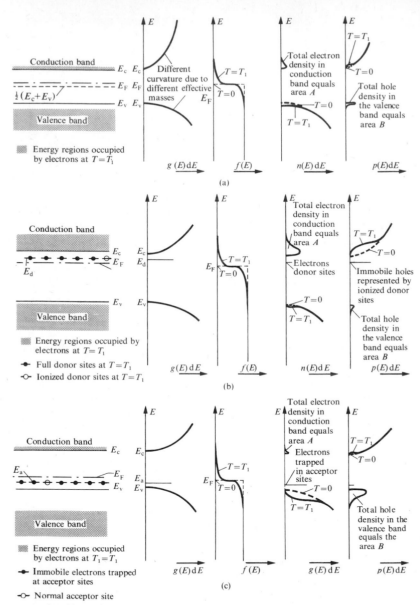

FIG. 5.6. The band model applied to three further types of semiconductor. (a) An intrinsic semiconductor with unequal effective masses for conduction electrons and holes. The different curvature of the $g(E)\,dE$ versus E plots near E_c and E_v, resulting from different effective masses, shifts E_F from the mid-point $\frac{1}{2}(E_c + E_v)$. (b) An n-type semiconductor. In the extrinsic regime the fermi energy is close to the conduction-band E_c. (c) A p-type semiconductor. In the extrinsic regime the fermi energy is close to the valence-band edge E_v.

relatively easily activated into this band. However, the resultant holes are localized and may only move by 'jumping' into the valence band. The energy required for this would be almost as much as that needed to break a bond of the host lattice. Such holes are thus doomed to a fixed and lonely existence. At temperatures sufficiently high to liberate donor electrons, but not to break many bonds of the host lattice, there will be more electrons than holes contributing to conduction. However, charge neutrality is maintained because the positively charged donor sites still exist even though they are immobile.

The position of the fermi level is thus determined so that the total number of conduction electrons (area A in Fig. 5.6(b)) equals the number of holes in the valence band (area B in Fig. 5.6(b)) *plus* the total number of ionized donor sites. Thus the fermi level lies close to the edge of the conduction band. Similarly, a high density of ionized acceptor sites leads to a fermi level which is close to the valence band edge (Fig. 5.6(c)).

Materials containing artificially introduced impurities are said to be *doped* and the impurities are termed *dopants*. Donor sites lead to a surfeit of mobile negative charges and materials with an excess of donor impurities are called n-type (negative type) materials. Similarly, an excess of acceptor sites gives rise to p-type (positive type) materials. Since the nature of the dominant current carriers is determined by externally introduced impurities, the resultant conduction is known as *extrinsic conduction*.

5.2.3. Carrier generation in semiconductors

The most characteristic property of a semiconductor is that before conduction may occur, a definite amount of energy must be supplied to excite carriers into a conducting state. In a metal the number of carriers is fixed and the temperature dependence of the electrical conductivity arises chiefly from the decrease in mobility with increasing temperature which results from scattering mechanisms. However, in a semiconductor the number of carriers is not fixed, but increases with increasing temperature, leading to a large increase of electrical conductivity with temperature. Let us look briefly at the mode of this change.

As an example, consider a n-type doped semiconductor. At absolute zero of temperature it would behave as a perfect insulator. Increasing the temperature imparts energy of the order of $k_0 T$ to the lattice. This is used initially to release electrons from donor sites. The number of electrons freed (n_d) depends on both the ratio of $k_0 T$ to the binding energy of the electrons at the donor sites and on the number of empty levels available in the conduction band.

If the temperature is further increased, sufficient energy may become available to rupture bonds of the host lattice. This itself results in the formation of pairs of electrons and holes which become available for conduction. This process of carrier activation is also very temperature dependent and obeys the equation

$$n_i = p_i = K \exp(-E_g/2k_0 T), \tag{5.8}$$

where K is a quantity which does not vary markedly with temperature, and n_i and p_i are the intrinsic concentrations of electrons and holes produced respectively. E_g is the bonding energy (or band gap $E_c - E_v$) and k_0 is Boltzmann's constant. The parameter K takes account of the availability of energy levels into which the carriers may be activated. At high temperatures, the intrinsic carrier concentrations n_i and p_i are much greater than n_d and the material behaves increasingly as an intrinsic semiconductor with E_F near the centre of the band gap.

Carrier generation provides a means whereby many physical properties of semiconductors may be explained. Firstly the rapid increase in electrical conductivity with increasing temperature in the intrinsic or extrinsic ranges results from the increasing carrier density. In contrast, the electrical conductivity of *metals* decreases slowly with temperature as a result of decreasing mobility. In semiconductors, carrier generation swamps the effect of the mobility.

Secondly, semiconductors are optically transparent at certain visible or near-visible wavelengths whereas metals are opaque. Modern theory pictures light as transmitted in packets of energy called photons, the energy of which is inversely proportional to the wavelength of the light. Now when light of an appropriate wavelength is incident on a semiconductor the photons may possess sufficient energy to generate electron–hole pairs. Thus electromagnetic energy is absorbed.† For larger wavelengths the photons are incapable of breaking bonds and are transmitted by the material.

It is illustrative to consider the optical transmission of two common forms of carbon. For diamond, the binding corresponds to a critical wavelength of about 170 nm (in the ultraviolet region of the electromagnetic spectrum). Hence visible wavelengths are transmitted. In graphite, however, which is also composed of carbon atoms, the negligible band gap in certain directions of the crystal prevents the material from being transparent to visible light.

Although not strictly appropriate to carrier activation, it is convenient at this juncture to consider another property in which semiconductors and metals differ. This is the thermal conductivity. In our discussion of metals in section 5.1.4, we interrelated the electrical and thermal conductivities. Since insulators are poor electrical conductors, we might expect them to be poor thermal conductors. This is not the case. For example, diamond, which we were discussing above, is one of the best thermal conductors available. The reason for this is that in insulators and semiconductors, it is not predominantly the electrons but the lattice which is responsible for thermal conduction. The lattice waves introduced in Chapter 2 are able to transport heat through the crystal.

---

† In practice the 'optical band gap' may differ from the 'thermal band gap'. The interested reader is referred to the more advanced texts in the bibliography.

5.3. Electron motion in amorphous materials

We recall that the band picture may be applied to amorphous materials (Chapter 4). If so, what effect does the disorder, which is characteristic of these materials, have on the electron motion? In practice, it is very marked. We shall consider why.

In an ideal crystal, conduction electrons may move freely. Such freedom is a consequence of the periodic structure of the ideal lattice. If one pictures the conduction electrons collectively as a distribution of negative charge within the crystal, then the electron motions can be considered as time-dependent fluctuations in this charge density. Such fluctuations are equivalent to a wave motion within the material lattice. In Chapter 2 we discussed a multiplicity of defects in a crystal which may contribute to the electrical resistance of a material by scattering the conduction electrons. A simple analogy is the deflection of water waves by stones or reeds which project through the surface of a pond. Such scattering processes oppose the force due to an applied electric field which tries continuously to accelerate electrons and holes. The effect of scattering is estimated in terms of the average distance an electron or hole may travel between scattering events. This distance is called the *mean free path* (l); in the ideal crystal, it is infinitely large. With many real crystals, it is of the order of 1000 atomic separations as this is the distance over which periodicity is normally maintained. In amorphous materials, however, l is very much smaller than this. For example, in liquid metals l is larger, but not much larger than a single interatomic spacing, whilst in liquid semiconductors and many conducting glasses, l appears to be less than an interatomic spacing. In these cases, it becomes difficult to imagine how electric charges can flow through the lattice at all!

This problem has led to speculation on the nature of electron motion in the amorphous state. One idea is that the electrons move by discrete *hops* between atomic-like localized levels in the material. Vibrational energy generally provides the stimulus for the hops, between which there may be a considerable time interval. An alternative approach considers the material as containing regions within which electrons may flow with relative ease, the regions being separated by barriers which hinder the free exchange of the electrons. The exact nature of electron transport in amorphous materials is a topic of active, current research and has not yet been resolved.

5.4. Crystalline devices

From the foregoing sections of this chapter, it is seen that different materials can have, or can be manipulated so as to have, a variety of electrical properties. These materials can be used to make a variety of electronic devices. In the rest of this chapter, we shall briefly consider the basis of some of these devices.

The most straightforward device is the *thermistor*. On p. 65, we noted that the electrical conductivity of semiconductors usually increases with temperature, and that the temperature coefficient of resistance is large. The thermistor (or thermally sensitive resistor) consists of a bead or disc of material to which electrical connection is made. It is used as a resistance thermometer. Also, when placed in series with other electrical components, its negative temperature coefficient of resistance compensates the positive coefficient of the other components and hence stabilizes the circuit against warming-up effects and other temperature changes.

Generally, the term 'device' is used for a solid block within which there are one or more junctions between two or more types of material. Such interfaces confer specific properties on the conductivity of the newly formed sample. For example, devices are produced which amplify electric currents, or allow current to flow in one direction only, or even convert optical energy to electrical energy and vice versa. There are two predominant factors which govern the behaviour of these devices. One is the nature of the band structure of the materials used to form the junction and the other is the distribution of electrons within these bands. Now, the formation of a junction involves the joining together of materials which, in isolation, would have different fermi energies. Immediately such materials are joined, carrier motion occurs in an attempt to maintain a constant fermi energy within the composite material of the device. Such motion leads to a redistribution of charge within the device to form an equilibrium state with a constant E_F. The operation of the device involves applying a suitable potential difference (p.d.) across the junction. This disturbs the charge distribution and artificially separates the fermi level on either side of the interface. The properties of a particular device are thus a reflection of how the junction attempts to thwart the applied p.d. and re-establish the equilibrium state. Specific devices thus involve choosing materials which have appropriate band structures and suitably occupied levels, so as to achieve the required equilibrium state upon union.

In practice, two types of junction are common. The first consists of a metal with a semiconductor. The fermi level in the metal is a fixed energy which can be varied only by changing the metal. With the semiconductor, at a given temperature, E_F may, within certain limits, be preselected by the deliberate addition of impurities. These devices will be considered in the next section. The second type of device is based on the fusion of two or more semiconducting components, for which the specifically added impurities are different in type and density in each component. The majority of crystalline devices are of this type.

5.4.1. *Metal–semiconductor junctions*

One of the earliest types of device consisted of a metal–semiconductor junction. The semiconductor could be either n- or p-type. Let us concentrate on a metal to n-type junction. Now, the n-type material has, in the extrinsic region, a fermi level near the conduction band. A metal is chosen such that E_F

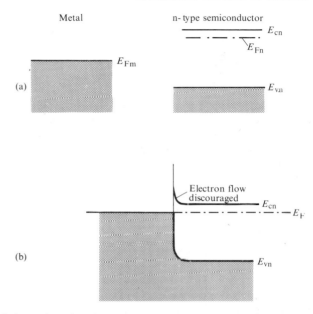

FIG. 5.7. A metal–semiconductor junction. (a) Band diagrams for the isolated metal and semiconductor. (b) Band diagram for the formed junction. E_{Fn} and E_{Fm} are respectively the fermi levels for the isolated n-type semiconductor and the metal, and E_{vn} and E_{cn} are respectively the valence- and conduction-band edges of the n-type semiconductor.

in the metal is greater than E_v but less than E_c for the semiconductor (Fig. 5.7(a)). When the materials are brought together, the fermi energy becomes constant throughout the device. This may be visualized as follows: as the union is made, those electrons which are near the junction in the semiconductor and which are in the relatively high energy states, lose energy by falling into available lower energy states in the metal. The metal acquires a net negative charge. Loss of electrons near the junction simultaneously exposes the positively charged empty donor sites in the semiconductor, resulting in a net positive charge in this material. Eventually the charge distribution becomes sufficiently large to prevent further electron flow across the boundary, and the device reaches a state of equilibrium (Fig. 5.7(b)). The curvature of the conduction-band edge signifies that it is energetically unfavourable for conduction electrons to progressively approach the metal.

The operating procedure involves perturbing this equilibrium state by applying an external p.d. across the junction. This separates the fermi energies on either side of the junction. Now, if the p.d. is applied in such a direction as to make the metal positive, electrons are encouraged to flow into the metal from the semiconductor. However, if the p.d. is applied so that the metal is made negative, the external field reinforces the internal field, and strongly prohibits electron flow. Thus, in one sense of bias† conduction across the junction is enhanced, while in the other it is resisted. The device allows current to flow in one direction only and is called a *rectifier*.

A major use of rectifiers is in the conversion of alternating current to direct current. In common circuitry, devices have metal contacts and interconnectors such as terminals and wires. It would thus be tragic if all metal contacts to semiconducting devices and components actually rectified. We recall that the rectifying character arises because the individual components are chosen to have different fermi energies. If materials are selected in which, even in isolation, E_F is identical, then upon union no rectifying property is evident. Such junctions can pass current equally in either direction upon suitable biasing. They are termed *ohmic contacts*. In practice, they are difficult to produce.

5.4.2. *The p–n junction diode*

The best-known and simplest interface device is the p–n junction or diode. We shall picture all external contacts as ohmic and concentrate on the behaviour of the interface. The interface marks a boundary in the semiconductor host lattice between n-type and p-type doped regions. In isolation (and in the extrinsic range of conduction), we would expect the fermi energy in the n-type region to be near the conduction-band edge and that in the p-type region to be near the valence-band edge. However, across the interface the fermi level must, under equilibrium conditions, be constant.

Upon forming the interface, highly energetic electrons in the n-type region, which are near the interface, fall into lower energy levels in the p-type region. This exposes positively charged, empty donor sites in the n-type material. Similarly, holes from the p-type region drop into lower 'hole states' in the n-type region, exposing negatively charged empty acceptor states in the p-material. Thus a charge distribution is built up near to, and on either side of, the junction. In equilibrium this barrier is just sufficient to oppose a net electron flow from the n-type to the p-type regions and a net hole flow from p-type to n-type material. Hence the energy band picture has the curvature shown in Fig. 5.8(a).

We use the device by biasing the junction into a non-equilibrium condition. If the n-type region is made negative, the biasing is called *forward*. Then, electron flow is encouraged from the n-type to the p-type material, as is hole flow from the p- to the n-type material (Fig. 5.8(b)). Since both regions are highly doped, plenty of carriers are available and the resistance of the device is fairly low. However, if the n-type region is made positive (so-called *reverse*

---

† We use the term bias to indicate the polarity of the electric field applied to the device under operating conditions.

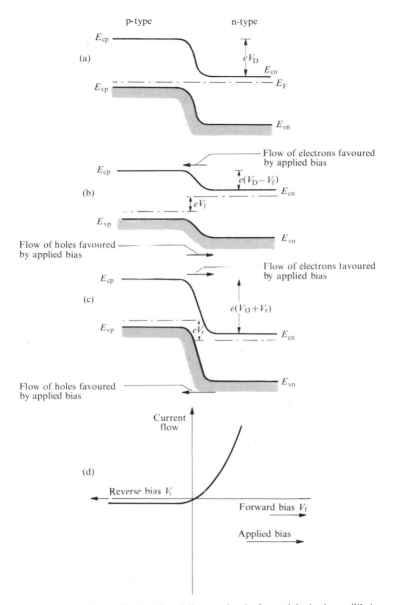

FIG. 5.8. Semiconductor diode (a) Band diagram for the formed device in equilibrium. (b) Band structure under forward bias. (c) Band structures under reverse bias. (d) The resulting current–voltage relationship for both types of bias. E_c and E_v are respectively the conduction band and valence band edges. Subscripts n and p indicate parameters for the n- and p-type material respectively. V_D is the internal p.d. set up by the exposed donor and acceptor sites at equilibrium.

bias), then the only possible conduction is by holes moving from the n- to the p-material and electrons travelling from the p- to the n-type substance (Fig. 5.8(c)). The electron density in the p-type region and the hole density in the n-region arise from intrinsic carrier generation, formed by the rupture of bonds in the host lattice. The density of such carriers will be low. Hence conduction under reverse bias is small, and the device resistance is high. The current versus voltage curve for the p–n junction is illustated in Fig. 5.8(d), whence we see its rectifying behaviour.

Another feature of the diode is that it can have capacitor-like properties. External fields reinforce, or partially offset, the internal barrier due to the charge distribution. In fact they alter the charge distribution near the junction by sweeping carriers to or from the junction. This means that the device can store charge, and hence act like a capacitor.

A third, and increasingly utilized property of the junction diode is its ability to convert optical energy into electrical energy. Earlier in this chapter, we pointed out that optical energy can be used to break bonds and generate carriers. Thus, if we shine light of a suitable wavelength on to a suitable junction, we may generate holes and electrons. The internal field across the junction rapidly sweeps these carriers into the p- and n-type regions respectively. Thus a current flows and optical energy is converted into electrical energy. Further, the intensity of incident light governs the number of broken bonds and hence the size of the current obtained. Also diodes sensitive to different regions of the spectrum may be produced by using different materials for the host lattice. Such optical detectors are called *photodiodes*.

In certain diodes it is possible to reverse the above effect and generate electromagnetic energy from electrical energy. We call such devices semiconductor *lamps*. By biasing in the forward direction, holes and electrons are injected into the n-type and p-type regions respectively. They recombine and emit light of a characteristic wavelength. This wavelength depends upon the bonding within the lattice, and is generally in the infrared region of the spectrum. However, careful choice of materials has produced lamps which emit visible light. Modern work has produced not just lamps but *lasers*, which can be made to give directional, intense, and coherent light. We will discuss lasers, together with the equally fascinating *tunnel diode*, in Chapter 6.

5.4.3. *The junction transistor*

The simplest multi-interface device is the transistor. It is formed by creating two p–n junctions arranged back to back. Either an n–p–n or p–n–p combination is possible. We will consider the former, and see how it may be used as an amplifier. One n-type region is termed the *emitter*, the central p-type region, which is made extremely thin, is called the *base*, and the remaining n-type region is termed the *collector*. The donor density in the emitter region is made larger than the acceptor density in the base region. In equilibrium, the fermi energy is constant throughout the device and charge distributions are established around the interfaces. When in operation, biases are applied which disrupt the equilibrium (Fig. 5.9). The emitter–base interface is biased in the forward direction (emitter negative, base positive) and the collector–base junction in the reverse sense (collector positive, base negative). This biasing encourages electrons to flow from the emitter into the base, and holes to flow from the base to the emitter. Since the emitter is more heavily doped than the base, the electron flow greatly exceeds the hole flow. Further, since

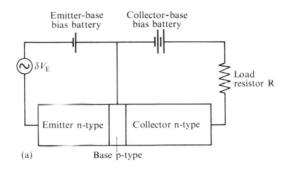

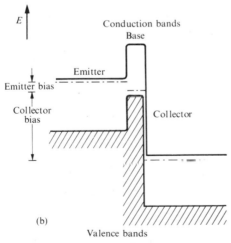

FIG. 5.9. The semiconductor transistor. (a) Circuit diagram. (b) Band diagram under operating conditions.

the base region is thin, the electrons entering it are immediately influenced by the reverse bias of the collector-base interface, with the result that they are whisked into the collector. This stolen electron current far exceeds the normal reverse-bias current of this interface.

Now suppose we vary the forward bias V_E of the emitter–base circuit by a small voltage δV_E. This leads to a change (δI_E) in the emitter current, the larger part of which consists of a change in electron current δI_{Ee}. Most of the electron density injected into the base region is captured by the collector. Hence the change in collector current $\delta I_C = \alpha \delta I_{Ee}$ where α is usually approximately 0·98. Hence the change δI_{Ee} leads to a change in collector current δI_C. This change in current produces a voltage change δV_R across the resistor of resistance R in the external collector circuit (Fig. 5.9). Clearly

$$\delta V_R = R\delta I_C = \alpha R(\delta I_{Ee})$$

whence the amplification is given by

$$\frac{\delta V_R}{\delta V_E} = \frac{\alpha R}{(\delta V_E/\delta I_{Ee})}.$$

If amplification is to occur we require R to be large with respect to the emitter–base internal resistance $\delta V_E/\delta I_{Ee}$. However, if R is too large, then the base–collector bias would fall largely across R rather than across the interface. We require $\delta V_E/\delta I_{Ee} < R < \delta V_C/\delta I_C$ which is clearly possible since we know $\delta V_E/\delta I_{Ee}$ to be small and $\delta V_C/\delta I_C$ to be large. Typically for $\delta V_E/\delta I_{Ee} = 50\Omega$, $\delta V_C/\delta I_C \approx 10^6\ \Omega$ and $R = 35\,000\ \Omega$ we have $\delta V_R/\delta V_E = 700$. This is a common value for the amplification in such devices.

Many other multi-interface devices are available with a whole selection of interesting properties. We refer the interested reader to the bibliography for details.

5.5. Amorphous devices

The operation and design of crystalline devices depends on the sensitivity of the electrical properties of crystalline semiconductors to small densities of added impurities. The aperiodic structure of amorphous semiconductors suggests that the influence of impurities may be small. In fact it is only recently that amorphous devices have been prepared. The operation of these devices depends on the fact that the resistance of certain glasses changes by several orders of magnitude when the potential difference applied across them exceeds a certain critical value. Quite why the resistance changes is not at present completely understood. However, the existence of two resistance states and the ability to switch from one state to another has suggested possible uses as 'on–off switches' in computers.

6. Quantum properties of materials

'Curiouser and curiouser'

LEWIS CARROLL *Alice in Wonderland*

6.1. Introduction

THROUGHOUT this book we have been considering how condensed matter responds when energy is exchanged with external sources. The important factor governing the response is the nature of the binding forces between the atoms of the medium. We have seen how the structure of the atoms effectively dictates the ways in which they can combine or bond together to form solids and liquids. Furthermore, it was shown that the bond concept, which simply reflects the nature and strength of the interatomic forces, can be developed into the band theory concept. Because these concepts have been used so successfully to account for the mechanical and electrical properties of condensed matter, one might think that they are comprehensive. In fact, they are not. For example, they cannot explain adequately why at very low temperatures, certain solids appear to have infinitely high electrical conductivity (i.e. effectively zero direct-current resistance) or why liquid helium at temperatures below 2·2 kelvin has properties consistent with an incredibly low viscosity. To account for these and other interesting phenomena, some of which are now being exploited commercially, we must return to a consideration of the basic properties of atoms themselves.

6.2. The quantum approach

In Chapters 1 and 2, the electronic structure of atoms is pictured in terms of discrete energy levels. If any external factor is to influence the atom, then it must impart (or subtract) one of a number of definite amounts (called *quanta*) of energy, corresponding to the excitation of the electrons into new energy levels. This picture of the atom is referred to as the *quantum model*. It may be more easily visualized with the help of an analogy. Many readers will be familiar with the common radio-receiver. Generally, a control is provided for tuning the receiver to a particular radio station. As the correct radio-wavelength is approached, the intensity of the received signal increases rapidly. Once past this wavelength, the signal decreases rapidly. At the optimum condition, your favourite programme can be heard clearly and loudly. Now, a modern radio-set receives more than one station, so that one may plot a response curve of the intensity of the received signal against wavelength (Fig. 6.1(a)). Nowadays, many homes and cars are equipped with radio-sets

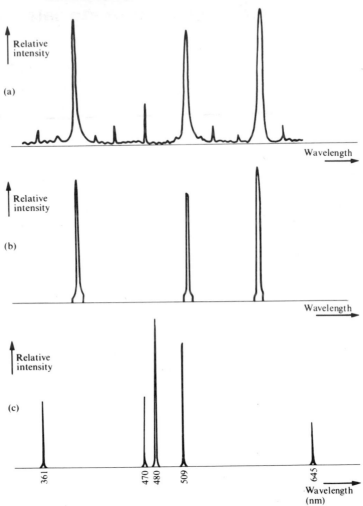

FIG. 6.1. Plots of relative intensity versus wavelength for: (a) A conventional radio receiver, (b) a push-button tuned radio-receiver, (c) the emission from a cadmium atom.

which have push-button tuning. Here, one selects a programme simply by pressing the correct button. If one plots a response curve for such a radio receiver one only observes a finite number of peaks determined by the number of selector buttons (Fig. 6.1(b)). .The quantum picture of the atom can be thought of as equivalent to this push-button system as it is limited to energy exchanges which correspond to discrete 'wavelength channels' (Fig. 6.1(c)).

Hitherto, we have considered the electron as if it were a particle. In the mathematical description of the quantum model, the electron is also pictured as a 'wave'. This dual view of electrons is one of the more difficult concepts of modern physics. Both representations are needed for a full explanation of many observed effects. For an electron in an atom, the wave will be a standing wave. To relate the two representations of the electron we must equate the 'intensity' of the standing wave with the probability distribution of the particle. Such probability distributions represent positions in space where the electrons are most likely to be found. They are the atomic orbitals to which we have often referred. Each electron has a charge associated with it. Hence, occupied atomic orbitals may be loosely viewed as charge distributions in space.

Absorption or emission of energy occurs when, on 'tuning' the atom the spatial distribution of charge changes from one discrete form to another. However, the atom resembles the push-button tuned system in that only a certain number of changes are permitted: the allowed and forbidden transitions are governed by strict rules called *selection rules*. Much as we often refer to radio-programmes by name, so a nomenclature has been devised to describe the absorption and emission of energy from atoms. However, whereas the name of a radio-station is often trivial, the nomenclature for atomic transitions provides definite information on the levels between which the energy changes occur. For this reason each atomic orbital has been associated with four quantum numbers. The principal quantum number n determines essentially the distance from the nucleus at which the maximum electronic charge density occurs. The second, or so called orbital quantum number l chiefly governs the shape, and the third, or magnetic quantum number m_l, the orientation of the atomic orbital. The fourth and final number m_s is the spin quantum number. This limits the occupation of each atomic level to a maximum of two electrons. The quantum numbers may only possess certain numerical values (Table 6.1). Each electron in an energy level of a given atom has a unique set of quantum numbers.

TABLE 6.1

Quantum-number nomenclature for atomic energy states.

Quantum numbers are numbers which may only possess certain specific values:
  n (the principal quantum number) may take any integral value from 1 upwards.
  l (the orbital quantum number) may, for each value of n, assume integral values from 0 up to $(n-1)$.
  m_l (the magnetic quantum number) has, for each value of n and l, possible positive and negative integral values in the range $-l, -(l-1), \ldots, 0, \ldots, (l-1), l$.
  m_s (the spin quantum number) may, for each value of n, l, and m_l, have the values $+\frac{1}{2}$ or $-\frac{1}{2}$.
Thus each unique set of values of n, l and m_l corresponds to an atomic energy level which may accommodate up to *two* electrons corresponding to the two values of m_s.

Table 6.1—*continued*

Quantum Numbers				Type of Atomic Orbital	Number of Atomic Orbitals	Type of Energy Shell	Maximum Number of Electrons Allowed
n	l	m_l	m_s				
1	0	0	$\pm\frac{1}{2}$	s-orbital	1	K	2
	0	0	$\pm\frac{1}{2}$	s-orbital	1		
2	1	-1 0 $+1$	$\pm\frac{1}{2}$ $\pm\frac{1}{2}$ $\pm\frac{1}{2}$	p-orbitals	3	L	8
	0	0	$\pm\frac{1}{2}$	s-orbital	1		
	1	-1 0 $+1$	$\pm\frac{1}{2}$ $\pm\frac{1}{2}$ $\pm\frac{1}{2}$	p-orbitals	3		
3	2	-2 -1 0 $+1$ $+2$	$\pm\frac{1}{2}$ $\pm\frac{1}{2}$ $\pm\frac{1}{2}$ $\pm\frac{1}{2}$ $\pm\frac{1}{2}$	d-orbitals	5	M	18
	0	0	$\pm\frac{1}{2}$	s-orbital	1		
	1	-1 0 $+1$	$\pm\frac{1}{2}$ $\pm\frac{1}{2}$ $\pm\frac{1}{2}$	p-orbitals	3		
	2	-2 -1 0 $+1$ $+2$	$\pm\frac{1}{2}$ $\pm\frac{1}{2}$ $\pm\frac{1}{2}$ $\pm\frac{1}{2}$ $\pm\frac{1}{2}$	d-orbitals	5		
4	3	-3 -2 -1 0 $+1$ $+2$ $+3$	$\pm\frac{1}{2}$ $\pm\frac{1}{2}$ $\pm\frac{1}{2}$ $\pm\frac{1}{2}$ $\pm\frac{1}{2}$ $\pm\frac{1}{2}$ $\pm\frac{1}{2}$	f-orbitals	7	N	32

etc.

However, the quantum model is not limited to electrons. In principle, any particle may be represented as a wave motion so that one may use this approach to calculate the energy and hence the properties of the particle. Such calculations are the realm of *quantum mechanics*.

Consider a simple, stable system consisting of two particles of the same type localized around points a distance r apart (Fig. 6.2(a)).† Since each is amenable to a wave representation, so is the composite pair. Using the wave concept, we may calculate the possible energy levels and their associated probability distributions for the system as a whole. Fig. 6.2(b) represents one such distribution for the simple one-dimensional system in question. The distribution function has maxima at the probable sites of the 'particles'. Now, if the particles are of the same type, then quantum mechanically they are said to be *identical* and the energy levels and their accompanying distributions are insensitive to an interchange of their positions. An important function is obtained by taking the square root of the probability distribution. It is called the *wave function* of the system, and is given the symbol Ψ.

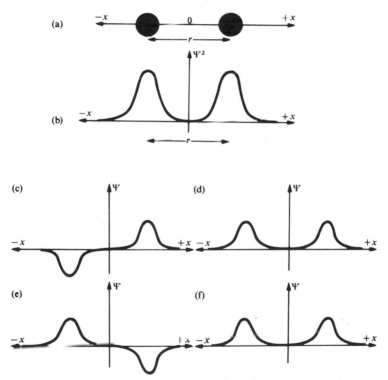

FIG. 6.2. Particle–wave duality. (a) A system of two identical particles in a given energy level. (b) The probability distribution Ψ^2 associated with that energy level. (c) and (e) Possible wave functions Ψ for a pair of fermions. (d) and (f) Possible wave functions Ψ for a pair of bosons.

† The arguments in this section are not rigorous, but give some insight into the consequences of identical particles in a quantum mechanical system.

Whereas the probability distribution, Ψ^2, may not be sensitive to the interchange of identical particles, this is not necessarily true of the wave function. For example, two possible wave functions are represented in Fig. 6.2(c) and 6.2(d). Both are compatible with Ψ^2 shown in Fig. 6.2(b).

Consider firstly the wave function of Fig. 6.2(c), and imagine that the particles are interchanged. The probability distribution (Fig. 6.2(b)) remains unchanged, but the wave function of Fig. 6.2(e) replaces that of Fig. 6.2(c). Now these two wave functions are mathematically the same except for a difference of sign. When an interchange of the particles leads to a change of sign of Ψ, the wave function is said to be antisymmetric. Particles which are characterized by antisymmetric wave functions include electrons and the helium isotope ^{3}He. They obey a form of statistics called Fermi–Dirac statistics (see section 5.1.2) and are known as *fermions*. An important property is that no two fermions may possess the same four quantum numbers and hence exist in the same quantum energy state. The astute reader may ask why we have allowed energy levels to be occupied by two electrons in Chapters 4 and 5. The answer is that the atomic energy levels are defined by three finite quantum numbers n, l, and m_l, whilst the spin number, m_s, may have one of two values. Hence, two electrons may exist in a given level as long as each satisfies one of the two values for m_s.

For the wave function illustrated in Fig. 6.2(d), interchanging the particles does not affect Ψ as shown in Fig. 6.2(f). The wave function is said to be symmetric. Photons and the ^{4}He isotope are examples of this type of 'particle'. They obey a different type of statistics known as Bose–Einstein statistics and are thus called *bosons*. An important consequence of the symmetry of their wave functions is that any number of them may inhabit a given quantum energy state.

The difference between fermions and bosons is well illustrated by the properties of the helium isotopes, ^{3}He and ^{4}He. Whereas at temperatures below 2·2 kelvin, ^{4}He, a boson fluid, behaves as a superfluid (see section 6.3.4 below), ^{3}He, a fermion fluid, remains a normal fluid at similar temperatures.

6.3. Quantum effects revealed

We have explained that the quantum picture can be applied to all particles and hence to all materials. One implication of this is that energy is exchanged in real systems by means of a large number of finite, discrete jumps or exchanges as the energy of the composite particles changes between discrete energy levels. Yet if we kick a football into the air and it loses energy to the resistive air, we do not see the ball slowing down in jerks. In this situation, the total energy changes are extremely large in comparison with the separation of the individual energy levels. We consider the overall motion and energy exchange of the ball: this typifies the classical approach. However, if the separation of individual energy levels is large or comparable with the total energy changes involved then quantum properties become evident. It is such cases that we shall deal with in the following sections.

6.3.1. *The tunnel effect*

Consider an electron whose motion is inhibited by an energy barrier. An example is an electron bound to an atom in a lattice. From what we have said previously in this book one might imagine that the only hope the electron has of moving through the lattice is that sufficient energy be provided to liberate it from its captive state.

However, the wave picture is capable of explaining an otherwise astonishing aspect of electron motion. Under certain circumstances, it is found that the electron may penetrate this barrier without having the energy to surmount it. The electron is said to *tunnel* through, rather than be activated over, the barrier. On a classical consideration, this would appear to be impossible, yet the quantum model provides us with an explanation. If we think of the probability distribution Ψ^2 associated with the electron and remember that it expresses the probability of the electron occupying a certain position in space (recall Fig. 6.2), we see that, provided the energy barrier is geometrically very thin, Ψ^2 may extend through the barrier, thus implying a small but finite probability for the electron to exist beyond the barrier. The effect is represented diagrammatically in Fig. 6.3.

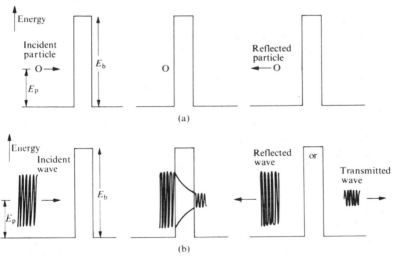

FIG. 6.3. Tunnelling through energy barriers. (a) Particle concept. A particle of energy E_p has insufficient energy to surmount an energy barrier E_b and is reflected. (b) The wave associated with the particle has a certain probability of passing through the barrier if this is thin enough. Hence the wave *may* be transmitted or reflected. Both diagrams are highly schematic.

Not only is the tunnelling of electrons observed experimentally, but it is used in commercial devices. One such device is the tunnel diode. This is a p–n junction (section 5.4.2) in which the two regions are heavily doped, so that, in equilibrium, the fermi level lies within the conduction band in the n-type region and within the valence band in the p-type region. The width of the junction is small (Fig. 6.4(b)). Under reverse bias, electrons may tunnel across the forbidden energy gap from the valence band of the p-type region to the conduction band of the n-type region (Fig. 6.4(a)). At low forward bias, the electrons tunnel in the opposite sense (Fig. 6.4(c)). As the forward bias is increased there is greater incentive for electrons to tunnel but, since the n-type conduction-band edge E_{cn} approaches the p-type valence-band edge E_{vp}, fewer empty levels are available to tunnel into. Hence, with increasing forward bias, the tunnelling current goes through a maximum and falls to zero when $E_{cn} = E_{vp}$. At higher bias voltages, only the normal thermal current flows in the junction (Fig. 6.4(d)). The unusual I versus V characteristic so obtained (Fig. 6.4(e)) should be compared with the normal diode characteristic, shown in Fig. 5.8. The tunnel-diode characteristic suggests that it might be used as an oscillator or as an electronic switch.

We have explained tunnelling with reference to electrons. It should be noted that it is a general wave phenomenon which is not confined to electrons alone.

6.3.2. Superconductivity

If a metal is cooled to lower and lower temperatures, its direct-current resistance decreases. With certain substances, a temperature is reached below which this electrical resistance suddenly vanishes. This astonishing, sudden absence of any impediment to the flow of electrical current has earned the phenomenon the name of superconductivity. It has been observed in a number of metals, compounds, and alloys, and is very sensitive to impurities and lattice defects. When superconducting, there is no dissipation of 'electrical energy' in the form of heat so that any d.c. current induced into a sample continues unattenuated for an indefinite time. It has been reported that such a current, induced in a ring of lead, was still flowing more than a year later.

Superconductors also have fascinating magnetic properties. Under normal conditions, a metal sustains a magnetic field. Upon becoming a superconductor, the loss of d.c. resistance is accompanied by a loss of the ability to support a magnetic field, and the field is rejected by the superconductor (Fig. 6.5). This may be demonstrated by the captivating 'floating magnet' experiment. A bar magnet is placed within a metallic bowl and is found to rise and remain suspended in mid-air when the metallic bowl becomes superconducting. The magnetic flux due to the bar magnet is rejected by the superconductor. The result is a force between the magnet and the bowl which in equilibrium, balances the gravitational force acting on the magnet. The temperature (T_c) at which these strange properties suddenly arise is called the

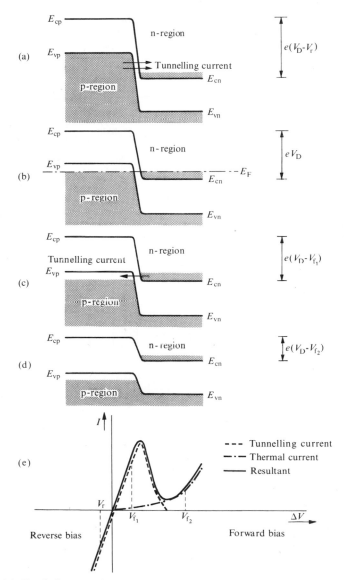

FIG. 6.4. Band diagrams for a tunnel diode. (a) Under reverse bias V_r showing tunnelling of electrons from the valence band of the p-type region into the conduction band of the n-type region. (b) Under equilibrium conditions. (c) Under low forward bias V_{f_1} showing tunnelling of electrons from the conduction band of the n-type region into the valence band of the p-type region. (d) Under high forward bias V_{f_2} whence a normal thermal current flows in the junction. (e) The current (I) versus voltage (ΔV) characteristic obtained for the device.

E_{cp} and E_{vp} are the conduction- and valence-band edges for the p-type region and E_{cn} and E_{vn} are equivalent energies for the n-type region.

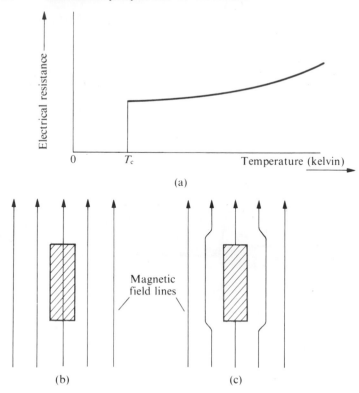

(a)

Magnetic
field lines

(b)   (c)

F IG. 6.5. The superconducting transition. (a) Schematic plot of electrical resistance
versus temperature for a metal showing the transition at the critical temperature T_c.
(b) Magnetic field entering a metal held at temperature $T > T_c$. (c) The expulsion of
magnetic field when T is lowered beneath T_c.

critical temperature. It is not the same critical temperature as that in section 1.2.
T_c is found to be characteristic of the particular metal involved, and furthermore shows a definite variation with the mass of different isotopes of the same
material. Some typical values of T_c are 7·2, 3·7, 1·2 and 0·9 kelvin for lead, tin,
aluminium, and zinc respectively, whilst two isotopes of tin with mass numbers
of 114 and 124 have T_c values of 3·81 and 3·65 kelvin respectively. Although no
structural change has been observed in the lattice at T_c, it is known experimentally that the system condenses to a lower energy state and that the
energy change involved is extremely small.

The experimental evidence presented theoreticians with a difficult problem,
and many years elapsed before a feasible explanation was given. Here we
outline how the quantum approach can give some indication of the origins of
the phenomenon.

The absence of a structural change and the small energy change involved (approximately 10^{-22} joules) suggest that the effect is related to some ordering of the conduction electrons rather than the atoms. This indicates that inter-action between the electrons is important. However, the experimental fact that T_c depends upon the mass of the isotopes for a given metal, implies that there is an electron interaction which occurs via the lattice. The proposed mechanism is that this interaction may under certain conditions be capable of overcoming the normal electrostatic repulsion, and results in a pairing of electrons of opposite momentum and spin. Such paired electrons are referred to as *Cooper pairs*. Now in the absence of a current flow, the net momentum of all Cooper pairs is identically zero. When a current is introduced into the superconductor, there is a movement of these pairs which represents a gain in momentum in the direction of current flow. All pairs share an equal increase in momentum. Now, it can be shown that if the velocity of a current carrier is less than that of the velocity of sound (s) within the lattice, then it cannot lose energy to, and hence be scattered by, lattice vibrations.† Thus, if the energy provided to generate the current is sufficiently small so as not to rupture the Cooper pairs, nor to make their velocity exceed s, then these pairs cannot be scattered and the electron pairs in the metal experience no resistance to motion. The material is then superconducting.

If the phenomenon depends simply on a question of electron-pair velocity, one might ask why all metals are not superconductors and why super-conductivity does not occur at higher temperatures. The answer to the first question is that Cooper-pair formation involves an interaction between electrons and the lattice,‡ and hence only those substances in which the electrons are sufficiently tightly bound to the lattice can become super-conductors. The second question can be answered as follows. At high tempera-tures, say T, the higher energy results from a distribution of momentum (and hence velocity) amongst the conduction electrons resulting in only a minute fraction of these electrons in the range $(E_F - k_0 T)$ to $(E_F + k_0 T)$ having veloci-ties less than s. Further, it is only at low temperatures that the motion of the electrons can be sufficiently correlated to provide *pairs* of electrons with the same net momentum. Hence, a significant number of Cooper pairs with velocities less than s is obtained only at a suitably low temperature.

The rejection of magnetic flux is also the result of the coordinated motion of electrons in the superconducting phase. Under normal conditions, an applied magnetic field influences the motion of electrons in a metal. This movement of charge leads to the formation of an additional, internal magnetic field which

---

† See, for example, the book by A. W. B. Taylor listed in the bibliography.

‡ Whereas lattice scattering involves an exchange of energy between electrons and the lattice in which energy and momentum are conserved, the electron–lattice interactions involved in Cooper-pair formation do not involve such an exchange. This fact is discussed in more advanced texts.

opposes the applied field. This is the origin of diamagnetism. Usually, the motion of the electrons leads to a partial cancellation of the opposing flux from various electrons. Under superconducting conditions, however, a well coordinated movement of electrons is favoured and a strong magnetic field is generated. This is of such a magnitude and direction as to cancel completely the effect of the applied flux density within the superconducting material.

6.3.3. *Josephson effects*

This section involves a combination of the ideas developed in the two foregoing sections. Let us consider a device consisting of a thin (about 10^{-8} m) strip of insulator mounted between two superconducting regions. The very nature of superconductivity indicates that the electrical resistance of this system must be due to the insulator. If a small potential difference ($\sim 10^{-6}$ volts) is applied across the arrangement, electrons can be induced to flow as the insulator is thin enough for the electrons to tunnel through it. The resulting current will only be due to the transport of those electrons which are *not* linked as Cooper pairs, as the pairs are effectively unable to tunnel through this thickness of barrier. The current therefore is very small. Incidentally, this arrangement provides a means of measuring the extremely small energy required for the formation of superconductivity: if the potential difference is increased, a value will be reached which delivers sufficient energy to rupture the Cooper pairs. The current suddenly increases as a flood of single electrons becomes available for the tunnelling process. The sharp change in the current versus voltage curve indicates the energy required to break up a Cooper pair, and so suggests the range of temperature over which superconductivity is likely to occur.

We return to our superconductor–insulator wafer. Josephson proposed that, if the wafer be made thin enough (about 10^{-9} m), then the Cooper pairs themselves would have a finite probability of penetrating (i.e. tunnelling through) the insulator. Now, Cooper pair mobility is the very essence of superconducting properties. Thus, a current of Cooper pairs can flow through the system without energy loss and without any voltage difference being either induced or continuously applied across the insulating layer. The system behaves as if it were a modified, single superconducting element. This specific ability of Cooper pairs to tunnel through a thin insulator is called the *d.c. Josephson effect*. If the current rises too high, the Cooper pairs generate enough kinetic energy to cause their own disruption into single electrons. These then move through the barrier with a finite resistance and the 'superconducting' properties across the junction are lost. One can see how such a property, in which the junction can switch from a zero to a finite resistance state can form the basis of a binary code for use in computer memory systems. Also, without giving details, it should be noted that this Josephson effect is strongly influenced by magnetic fields and has been used as a sensitive detector for small magnetic fields.

Josephson predicted a second effect. The existence of a small, finite d.c. potential difference across the insulator within a Josephson junction does not, of itself, prevent Cooper pairs tunnelling. It does, however, impart more energy to pairs on one side of the barrier. On tunnelling through the insulator, this difference in energy results in an oscillating supercurrent which emits electromagnetic radiation with a characteristic frequency, determined by the magnitude of the applied potential difference. This is the *a.c. Josephson effect*. It has been revealed in terms of discrete steps in the current versus voltage characteristic when the junction is activated and simultaneously irradiated with other (microwave) radiation. The steps occur when the applied microwave frequency, or a multiple of it, equals the frequency of the oscillatory supercurrent, and the two mix to give an additional direct current. This is the basis of current interest in the a.c. effect as a sensitive means of detecting radiation.

6.3.4. *Superfluidity*

Another example of the unmasking of quantum properties can be found in the properties of liquid helium† at very low temperatures. Helium is an element which may be cooled (under normal pressure) to the lowest temperatures available without solidifying. However, below 2·2 kelvin the liquid acquires a strange, almost magical, character. It is known as helium I above 2·2 kelvin and helium II below this temperature. We cannot hope to describe fully the peculiar behaviour of helium II but we shall indicate some of the more striking properties and hint at their theoretical interpretation.

Let us note some of the observable properties of liquid helium. Firstly, when helium I is cooled, evaporation under reduced pressure is indicated by bubbles forming at the walls of the container. On cooling below 2·2 kelvin, bubble formation ceases even though evaporation still occurs in the newly formed helium II. This quiescent boiling has been attributed to a remarkably high thermal conductivity for helium II leading to heat conduction away from the walls of the container. A second astonishing effect occurs when a beaker is partially lowered into a bath of helium II. The liquid is found to creep up the sides of the beaker, to climb over the lip of the vessel and to collect inside it. If the beaker is then removed from the bath, the helium returns to the bath by creeping back along the surface of the beaker (Fig. 6.6). In Chapter 3 it was mentioned that many liquids form films across solid surfaces. The rate of film formation depends upon the liquid viscosity. Most liquids have such a high viscosity that one does not notice any actual flow. The 'creeping' properties of helium II are explicable if this fluid has a remarkably low viscosity, and a low viscosity can be compatible with a high thermal conductivity *if* heat transport were by conduction currents within the fluid.

† In this section, references to helium refer to the isotope ^{4}He unless otherwise specified.

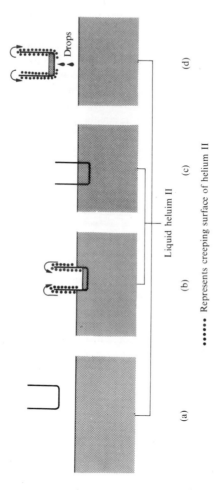

•••••• Represents creeping surface of helium II

FIG. 6.6. The creeping-film effect exhibited by liquid helium (^{4}He). (a) An empty beaker is lowered into a bath of helium II. (b) The helium II creeps up the sides of the beaker in an attempt to maintain equal levels of helium II inside and outside the beaker. (c) The desired equilibrium. (d) If the beaker is removed, the liquid helium creeps out of the beaker and drops back into the bath.

Unfortunately, viscosity measurements on helium II present an apparent paradox. Different methods of measurement yield wildly varying results for the viscosity coefficient. Methods which involve flow through a narrow slit or along a capillary tube do give a negligibly small value for the viscosity. On this observation, the fluid was dubbed a *superfluid*. On the other hand, measurements of the drag experienced by an object which was rotated or allowed to oscillate in the fluid gave much higher viscosity values, more consistent with a normal liquid. This discrepancy proved to be significant in accounting for the behaviour of helium II.

To explain the properties of helium II it is necessary to calculate the energy of a system of densely packed, interacting helium atoms. Fortunately, at such low temperatures, the energy of the system is close to that of the lowest (i.e. ground) energy level. On the quantum model, and using Bose–Einstein statistics, the higher energy levels are taken to represent a collective motion of the entire fluid rather than the independent motions of individual atoms. This makes the higher energy levels difficult to excite and few such excitations occur at low temperatures.

Conceptually it is useful to picture helium II as containing two components: one a 'superfluid' component representing the ground level and the other a 'normal fluid' component representing the excited levels. On such a model, the discrepancy in viscosity values can be explained in the following manner. Flow through a small gap is confined to the superfluid component which, since it represents the ground level, cannot lose energy and hence possesses zero viscosity. However, a body rotated within a bath of liquid helium II experiences a restraint which is transmitted from the walls of the containing vessel via the normal component of the fluid. An elegant experiment was devised to test this two-component picture. Helium II was filtered through a fine filter into a second container. Such filtration should remove superfluid from the first container, thereby increasing the concentration of the normal component in the remaining fluid. Such a separation should correspond to an increase in energy of the system in the first container. It was manifest experimentally as an increase in temperature of the fluid in this container.

It is interesting to compare superconductivity and superfluidity. The superfluid state and the Cooper pairs (i.e. two electrons forming a boson) are both zero-entropy states which can propagate without resistance. The normal fluid and the single electrons, formed by breaking up a Cooper pair, suffer scattering and generate resistance. The boson character is important in producing a sharp transition into the 'super' state. Earlier, we have contrasted ^{3}He and ^{4}He at temperatures just below 2·2 kelvin. At these temperatures, ^{3}He is a fermion fluid. If ^{3}He atoms could interact to form a boson fluid, they would show 'super' properties. Present research suggests that this may be possible at extremely low temperatures.

6.3.5. *Lamps and lasers*

In the previous sections we have considered how unusual properties may be revealed when the vibrational energy stored in a lattice is reduced at low temperatures to values not large compared with the separation of individual energy levels. In Chapter 5, we have already met cases where this condition can be satisfied even at room temperatures. This was in the consideration of insulators and semiconductors in which the separation of energy levels (the band gap) is large compared to the vibrational energy. For example, we saw how if one imparts electrical energy to certain semiconductor devices, the system may emit radiation of a very discrete wavelength. This corresponds to the emission of discrete quanta of energy due to electron transitions within the material. The resultant devices are called semiconductor lamps.

What would happen if one were to shine light on a semiconductor? It may be absorbed if the photons are associated with the appropriate quantum of energy and bonded 'valence' electrons within the semiconductor may be released from their sites and take part in conduction. This is called photo-conduction. There is, however, a second, albeit remote, possibility in certain circumstances. Consider a medium containing free electrons. If these can be persuaded to return to their original bonded state, they may liberate a certain amount of energy, which will correspond to a certain photon energy and wavelength. Now, if photons of this wavelength are incident on the material, they stimulate the free electrons to fall to their bonded condition and liberate a second set of photons of identical wavelength to the incident ones; but the latter have given no energy to the electrons and thus still exist. The material emits more photons than it received, and has thus acted as a light amplifier. Under normal conditions, this stimulated emission is highly improbable. In a forward biassed p–n junction operated under high current densities, injection of many electrons into the p-type region and many holes into the n-type region occurs. Under the correct conditions, such a device favours stimulated emission rather than absorption. Any photon produced in the region of the interface rapidly leads to the formation of two identical photons, travelling in the same direction, by the process mentioned above. These photons will in turn give rise to further photon generation. This process of **L**ight **A**mplification by **S**timulated **E**mission of **R**adiation is the principle behind the semiconductor **LASER**. The reader is referred to the bibliography for details of the construction, working, and uses, of lasers.

Bibliography

'Woe be to him that reads but one book'

GEORGE HERBERT Jacula prudentum

The following list is provided as a starting point for the reader who wishes to gain more complete information on the topics discussed in this book. Certain sections are covered by a host of books which can be found on the shelves of any scientific library; our brief selection does not necessarily imply the best. On certain other topics it is relatively difficult to find literature. In these cases we have deliberately provided a fuller list of references and marked them with the symbol *. The number of entries listed for the topics below are not intended to represent their relative importance.

General texts

A Scientific American Book (1967) *Materials*, W. H. Freeman and Co., San Francisco.

PASK, J. A. (1967) *An atomistic approach to the nature and properties of materials*, Wiley, New York.

RICE, F. O. and TELLER, E. (1949) *The structure of matter*, Wiley, New York.

ROSENTHAL, D. (1964) *Introduction to properties of materials*, Van Nostrand, Princeton.

Chapter 1

GOTTLIEB, M., GARBUNY, M., and EMMERICH, W. (1966) *Seven states of matter*, Walker and Co., New York.

Chapter 2

BUECHE, F. (1962) *Physical properties of polymers*, Interscience, New York.

CARTMELL, E. and FOWLES, G. W. A. (1961) *Valency and molecular structure*, Butterworths, London.

EVANS, R. C. (1964) *Introduction to crystal chemistry*, 2nd edn., Cambridge University Press, London.

*FERGASON, J. L. (1964, August) *Liquid crystals*, in *Scientific American*, p. 77.

FRANK, F. (1952) *Crystal growth and dislocations*, Adv. Phys. **1**, 91.

*GREENE, D. (1972, 11 May) *Promise of the superionic conductors*, in *New Scientist*, p. 321.

*HEILMEIER, G. H. (1970, April) *Liquid crystal display devices*, in *Scientific American*, p. 100.

JONES, G. O. (1956) *Glass*, Wiley, New York.

KITAIGORODSKIY, A. I. (1967) *Order and disorder in the world of atoms*, Longmans, London.

MOFFAT, W. G., PEARSALL, G. W. and WULFF, J. (1964) *The structure and properties of matter—I structure*, Wiley, New York.

PAULING, L. (1960) *The nature of the chemical bond*, 3rd edn., Cornell University Press.

SPICE, J. E. (1964) *Chemical binding and structure*, Pergamon Press, London.

Chapter 3

BIKERMAN, J. J. (1970) *Physical surfaces*, Academic Press, New York.
BURDON, R. S. (1949) *Surface tension and the spreading of liquids*, Cambridge University Press, London.
COTTRELL, A. H. (1964) *The mechanical properties of matter*, Wiley, London.
DUGDALE, D. S. (1968) *Elements of elasticity*, Pergamon Press, London.
GORDON, J. E. (1968) *The new science of strong materials*, Penguin, London.
HAYDEN, W., MOFFATT, W. G. and WULFF, J. (1964) *The structure and properties of matter—III Mechanical behaviour*, Wiley, New York.
SPENCER, G. C. (1968) *Introduction to plasticity*, Chapman and Hall, London.

Chapters 4 *and* 5

DEKKER, A. J. (1958) *Solid state physics*, Macmillan, London.
*DAVIES, E. A. (1971) *Amorphous semiconductors*, in *Endeavour*, 30, 55.
GOLDSMID, H. J. (1965) *The thermal properties of solids*, Routledge and Kegan Paul, London.
*HENISCH, H. K. (1969, November) *Amorphous semiconductor switches*, in *Scientific American*, p. 30.
*IOFFE, A. F. and REGEL, A. R. (1960) *Amorphous electronic conductors*, *Prog. Semicond.*, 4, 237.
IOFFE, A. F. (1960) *Physics of semiconductors*, Infosearch, London.
KITTEL, C. (1971) *Introduction to solid state physics*, 4th edn., Wiley, New York.
MORGAN, D. V. and HOWES, M. J. (1972) *Solid state electronic devices*, Wykeham Publications Ltd., London.
*OVSHINSKY, S. R. (1969) *Amorphous semiconductors*, in *Science Journal*, p. 73.
PEASE, L., ROSE, R. M., and WULFF, J. (1964) *The structures and properties of matter—IV Electronic properties*, Wiley, New York.
SPROULL, R. L. (1972, December) *The conduction of heat in solids*, in *Scientific American*, p. 288.

Chapter 6

BOCKHOFF, F. J. (1969) *Elements of quantum theory*, Addison-Wesley, London.
DARROW, K. K. (1952, March) *The quantum theory*, in *Scientific American*, p. 47.
HERMANN, A. (1971) *The genesis of quantum theory*, M.I.T. Press, London.
HOFFMANN, B. (1947) *The strange story of the quantum*, Penguin, London.
LENGYEL, B. A. (1971) *Lasers*, 2nd edn., Interscience, New York.
LIFSHITZ, E. M. (1958, June) *Superfluidity*, in *Scientific American*, p. 30.
LOTHIAN, G. F. (1963) *Electrons in atoms*, Butterworths, London.
MATTHEWS, P. T. (1963) *Introduction to quantum mechanics*, McGraw Hill, London.
MATTHIAS, B. T. (1957, November) *Superconductivity*, in *Scientific American*, p. 92.
MOTT, N. F. (1972) *Elementary quantum mechanics*, Wykeham Publications Ltd., London.
RIEF, F. (1960, November) *Superfluidity and quasi-particles*, in *Scientific American*, p. 138.
TAYLOR, A. W. B. (1970) *Superconductivity*, Wykeham Publications Ltd., London.
SCHAWLOW, A. L. (1963, July) *Advances in optical masers*, in *Scientific American*, p. 34.

Index

'Of course one has to have an Index. Authors themselves would prefer not
to have any. Having none would save trouble and compel reviewers to read
the whole book instead of just the Index.'

STEPHEN LEACOCK *Index: there is no Index*

The periodic table of the elements

Outer energy shell	Group: I	IIA	IIIA	IVA	VA	VIA	VIIA	VIIIA			IB	IIB	IIIB	IVB	VB	VIB	VIIB	O
K	H $1s\,1$																	He $1s\,2$
L	Li $2s\,1$	Be $2s\,2$											B $2s\,2p\;2\,1$	C $2s\,2p\;2\,2$	N $2s\,2p\;2\,3$	O $2s\,2p\;2\,4$	F $2s\,2p\;2\,5$	Ne $2s\,2p\;2\,6$
M	Na $3s\,1$	Mg $3s\,2$											Al $3s\,3p\;2\,1$	Si $3s\,3p\;2\,2$	P $3s\,3p\;2\,3$	S $3s\,3p\;2\,4$	Cl $3s\,3p\;2\,5$	Ar $3s\,3p\;2\,6$
N	K $4s\,1$	Ca $4s\,2$	Sc $3d\,4s\;1\,2$	Ti $3d\,4s\;2\,2$	V $3d\,4s\;3\,2$	Cr $3d\,4s\;5\,1$	Mn $3d\,4s\;5\,2$	Fe $3d\,4s\;6\,2$	Co $3d\,4s\;7\,2$	Ni $3d\,4s\;8\,2$	Cu $3d\,4s\;10\,1$	Zn $3d\,4s\;10\,2$	Ga $3d\,4s\,4p\;10\,2\,1$	Ge $3d\,4s\,4p\;10\,2\,2$	As $3d\,4s\,4p\;10\,2\,3$	Se $3d\,4s\,4p\;10\,2\,4$	Br $3d\,4s\,4p\;10\,2\,5$	Kr $3d\,4s\,4p\;10\,2\,6$
O	Rb $5s\,1$	Sr $5s\,2$	Y $4d\,5s\;1\,2$	Zr $4d\,5s\;2\,2$	Nb $4d\,5s\;4\,1$	Mo $4d\,5s\;5\,1$	Tc $4d\,5s\;5\,2$	Ru $4d\,5s\;7\,1$	Rh $4d\,5s\;8\,1$	Pd $4d\,5s\;10\,0$	Ag $4d\,5s\;10\,1$	Cd $4d\,5s\;10\,2$	In $4d\,5s\,5p\;10\,2\,1$	Sn $4d\,5s\,5p\;10\,2\,2$	Sb $4d\,5s\,5p\;10\,2\,3$	Te $4d\,5s\,5p\;10\,2\,4$	I $4d\,5s\,5p\;10\,2\,5$	Xe $5s\,5p\;2\,6$
P	Cs $6s\,1$	Ba $6s\,2$		Hf $5d\,6s\;2\,2$	Ta $5d\,6s\;3\,2$	W $5d\,6s\;4\,2$	Re $5d\,6s\;5\,2$	Os $5d\,6s\;6\,2$	Ir $5d\,6s\;7\,2$	Pt $5d\,6s\;9\,1$	Au $5d\,6s\;10\,1$	Hg $5d\,6s\;10\,2$	Tl $5d\,6s\,6p\;10\,2\,1$	Pb $5d\,6s\,6p\;10\,2\,2$	Bi $5d\,6s\,6p\;10\,2\,3$	Po $5d\,6s\,6p\;10\,2\,4$	At $5d\,6s\,6p\;10\,2\,5$	Rn $5d\,6s\,6p\;10\,2\,6$
Q	Fr $7s\,1$	Ra $7s\,2$																

Lanthanides

Element	4f	5d	6s
La	0	1	2
Ce	1	1	2
Pr	3	0	2
Nd	4	0	2
Pm	5	0	2
Sm	6	0	2
Eu	7	0	2
Gd	7	1	2
Tb	8	1	2
Dy	10	0	2
Ho	11	0	2
Er	12	0	2
Tm	13	0	2
Yb	14	0	2
Lu	14	1	2

Actinides

Element	5f	6d	7s
Ac		1	2
Th		2	2
Pa	2	1	2
U	3	1	2
Np	4	1	2
Pu	6	0	2
Am	7	0	2
Cm	7	1	2
Bk	8	1	2
Cf	10	0	2
Es	11	0	2
Fm	12	0	2
Md	13	0	2
No	14	0	2

Each box in the periodic table contains one element. The elements are represented by their conventional chemical symbols. The position of each box in the table indicates the period (horizontal row) and the group (vertical column) to which the element belongs. Each period is identified by the letter specifying the outer or valence shell of electrons.

We have not specified the complete electronic structure of each element. Rather, we have indicated the occupancy of those atomic levels whose occupancy changes on crossing a period of the table. Thus elements within a given period are presumed in addition to possess the electronic structure specified for elements at the ends of preceding periods.

We have named groups of atomic levels by specifying the value of n (the principal quantum number) and the type of orbital corresponding to values of l (the orbital quantum number).

Hence

$$3d \text{ implies } (n = 3, l = 2)$$

$$4s \text{ implies } (n = 4, l = 0) \text{ etc.}$$

Beneath each group of atomic levels are shown the number of electrons occupying these levels. It should be noted that the electron configurations of some elements are a topic of current debate.

Physical constants and conversion factors

Avogadro constant	L or N_A	6.022×10^{23} mol^{-1}
Bohr magneton	μ_B	9.274×10^{-24} J T^{-1}
Bohr radius	a_0	5.292×10^{-11} m
Boltzmann constant	k	1.381×10^{-23} J K^{-1}
charge of an electron	e	-1.602×10^{-19} C
Compton wavelength of electron	$\lambda_C = h/m_e c$	$= 2.426 \times 10^{-12}$ m
Faraday constant	F	9.649×10^4 C mol^{-1}
fine structure constant	$\alpha = \mu_0 e^2 c/2h$	$= 7.297 \times 10^{-3}$ ($\alpha^{-1} = 137.0$)
gas constant	R	8.314 J K^{-1} mol^{-1}
gravitational constant	G	6.673×10^{-11} N m^2 kg^{-2}
nuclear magneton	μ_N	5.051×10^{-27} J T^{-1}
permeability of a vacuum	μ_0	$4\pi \times 10^{-7}$ H m^{-1} exactly
permittivity of a vacuum	ϵ_0	8.854×10^{-12} F m^{-1} ($1/4\pi\epsilon_0 = 8.988 \times 10^9$ m F^{-1})
Planck constant	h	6.626×10^{-34} J s
(Planck constant)/2π	$\hbar$	1.055×10^{-34} J s $= 6.582 \times 10^{-16}$ eV s
rest mass of electron	m_e	9.110×10^{-31} kg $= 0.511$ MeV/c^2
rest mass of proton	m_p	1.673×10^{-27} kg $= 938.3$ MeV/c^2
Rydberg constant	$R_\infty = \mu_0^2 m_e e^4 c^3/8h^3$	$= 1.097 \times 10^7$ m^{-1}
speed of light in a vacuum	c	2.998×10^8 m s^{-1}
Stefan–Boltzmann constant	$\sigma = 2\pi^5 k^4/15h^3 c^2$	$= 5.670 \times 10^{-8}$ W m^{-2} K^{-4}
unified atomic mass unit (^{12}C)	u	1.661×10^{-27} kg $= 931.5$ MeV/c^2
wavelength of a 1 eV photon		1.243×10^{-6} m

1 Å $= 10^{-10}$ m; 1 dyne $= 10^{-5}$ N; 1 gauss (G) $= 10^{-4}$ tesla (T);
$0°$C $= 273.15$ K; 1 curie (Ci) $= 3.7 \times 10^{10}$ s^{-1};
1 J $= 10^7$ erg $= 6.241 \times 10^{18}$ eV; 1 eV $= 1.602 \times 10^{-19}$ J; 1 cal$_{th}$ $= 4.184$ J;
$\ln 10 = 2.303$; $\ln x = 2.303 \log x$; e $= 2.718$; $\log$ e $= 0.4343$; $\pi = 3.142$